# MILLER'S

## *Classic Motorcycles*

## PRICE GUIDE

MILLER'S CLASSIC MOTORCYCLES PRICE GUIDE 1996

Created and designed by
Miller's
The Cellars, High Street,
Tenterden, Kent, TN30 6BN
Tel: 01580 766411

Consultants: Judith & Martin Miller

General Editor: Valerie Lewis
Special Consultant: Mick Walker

Editorial and Production Co-ordinator: Sue Boyd
Editorial Assistants: Gillian Judd, Marion Rickman, Karen Taylor, Jo Wood
Production Assistants: Gillian Charles
Advertising Executive: Melinda Williams
Index compiled by: DD Editorial Services, Beccles
Design: Jody Taylor, Kari Reeves, Matthew Leppard

First published in Great Britain in 1995
by Miller's, an imprint of
Reed Consumer Books Limited,
Michelin House, 81 Fulham Road
London SW3 6RB
and Auckland, Melbourne, Singapore and Toronto

© 1995 Reed International Books Limited

A CIP catalogue record for this book is
available from the British Library

ISBN 1-85732-658-4

Illustrations by G. H. Graphics, St Leonard's-on Sea, E. Sussex.
Colour origination by Scantrans, Singapore
Printed and bound in England by William Clowes Ltd
Beccles and London

Miller's is a registered trademark of
Reed International Books Ltd

# MILLER'S
## *Classic*
## *Motorcycles*
# PRICE GUIDE

Consultants
## Judith and Martin Miller

General Editor
## Valerie Lewis

Special Consultant
## Mick Walker

# 1996
## Volume III

# ACKNOWLEDGEMENTS

*Miller's Publications would like to acknowledge the great assistance given by our consultants.*

| | |
|---|---|
| **Malcolm Barber** | Brooks, 81 Westside, London, SW4 9AY. |
| **Jim Gleave** | Atlantic Motorcycles, 20 Station Road, Twyford, Berks RG10 9NT |
| **John Newson** | Oxney Motorcycles, Rolvenden, Cranbrook, Kent TN17 4QA |
| **Jody Taylor** | 114 Pound Road, East Peckham, Tonbridge, Kent TN12 5BJ |
| **Brian Verrall** | Woodlands, Mill Lane, Lower Beeding, W. Sussex RH13 6PX |
| **Mick Walker** | 10 Barton Road, Wisbech, Cambs PE13 1LB |

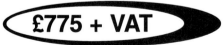
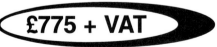

# KEY TO ILLUSTRATIONS

*Each illustration and descriptive caption is accompanied by a letter code. By referring to the following list of Auctioneers (denoted by \*) Dealers (•) and Clubs (§) the source of any item may be immediately determined. Inclusion in this edition no way constitutes or implies a contract or binding offer on the part of any of our contributors to supply or sell the goods illustrated, or similar articles, at the prices stated. Advertisers in this year's directory are denoted by †.*

*If you require a valuation for an item, it is advisable to check whether the dealer or specialist will carry out this service and if there is a charge. Please mention Miller's when making an enquiry. Having found a specialist who will carry out your valuation it is best to send a photograph and description of the item to the specialist together with a stamped addressed envelope for the reply. A valuation by telephone is not possible.*

*Most dealers are only too happy to help you with your enquiry, however, they are very busy people and consideration of the above points would be welcomed.*

| | | |
|---|---|---|
| **ABT** | • | A.B.T. Engineering, 18 Hollis Gardens, Luckington, Chippenham, Wiltshire, SN14 6NS. Tel: 01666 840275 |
| **ADT** | * | ADT Auctions Ltd, Classic & Historic Automobile Division, Blackbushe Airport, Blackwater, Camberley, Surrey, GU17 9LG. Tel: 01252 878555 |
| **ALC** | * | Alcocks, Wyeval House, 42 Bridge Street, Hereford, Hereford & Worcs., HR4 9DG. Tel: 01432 344322 |
| **AOC** | § | Ariel Owners Motorcycle Club, c/o Lester Grant, Cedar Cottage, Dicklow Cob, Lower Withington, Macclesfield, Cheshire, SK11 9EA. |
| **AT** | • | Tiernan, Andrew, Vintage & Classic Motorcycles, Old Railway Station, Station Road, Framlingham, Woodbridge, Suffolk, IP13 9EE. Tel: 01728 724321 |
| **ATF** | | Fletcher, A T (Enthusiast & Collector), Lancs. |
| **AtMC** | •† | Atlantic Motorcycles, 20 Station Rd, Twyford, Berkshire, RG10 9NT. Tel: 01734 342266 |
| **BCA** | | Beaulieu Cars Automobilia, Beaulieu, Hampshire, S042 7YE. Tel: 01590 612689 |
| **BCB** | •† | Bristol Classic Bikes, 17 Church Road, Bristol, Avon, BS5 9JJ. Tel: 0117 955 7762 |
| **BKS** | *† | Brooks, Robert (Auctioneers) Ltd, 81 Westside, London, SW4 9AY. Tel: 0171 228 8000 |
| **BLM** | •† | Bill Little Motorcycles, Oak Farm, Braydon, Swindon, Wilts., SN5 0AG. Tel: 01666 860577 |
| **BMM** | | Battlesbridge Motorcycle Museum, Muggeridge Farm, Maltings Road, Battlesbridge, Essex, SS11 7RF. Tel: 01268 769392/560866 |
| **BMW** | § | BMW Club, c/o John Lawes, (Vintage Secretary), Bowbury House, Kirk Langley, Ashbourne, Derbyshire, DE6 4NJ. Tel: 01332 824334 |
| **C(A)** | * | Christie's Australia Pty Ltd, Head Office – Melbourne, 1 Darling Street, South Yarra, Victoria, 3141. Tel: (03) 8204311 |
| **C.A.R.S.** | •† | C.A.R.S. (Classic Automobilia & Regalia Specialists), 4-4a Chapel Terrace Mews, Brighton, Sussex, BN2 1HU. Tel: 01273 601960 |
| **CMAN** | § | Christian Motorcycle Association North, c/o Mr A Sutton, 100 Low Bank Road, Ashton-in-Makerfield, Wigan, Gt. Manchester, WN4 9RZ. |
| **COB** | • | Cobwebs, 78 Northam Road, Southampton, Hampshire, SO14 0PB. Tel: 01703 227458 |
| **COEC** | § | Cotton Owners & Enthusiasts Club, c/o Peter Turner, Coombehayes, Sidmouth Road, Lyme Regis, Dorset, DT7 3EQ. |
| **CStC** | •† | Cake Street Classics, Bellview, Cake Street, Laxfield, Nr Woodbridge, Suffolk, IP13 8EW. Tel: 01986 798504 |
| **CVPG** | § | Chiltern Vehicle Preservation Group, Chiltern House, Aylesbury, Bucks., HP17 8BY. Tel: 01296 651283 |
| **DM** | •† | Mitchell, Don, & Company, 132 Saffron Road, Wigston, Leicestershire, LE18 4UP. Tel: 0116 277 7669 |
| **DOT** | § | Dot Owners Club, c/o Chris Black, 115 Lincoln Avenue, Clayton, Newcastle, ST5 3AR. |
| **EP** | * | Evans & Partridge, Agriculture House, High Street, Stockbridge, Hampshire, SO20 6HF. Tel: 01264 810702 |
| **Gen** | * | Gene Harris Antique Auction Centre Inc, 203 S. 18th Avenue, Marshalltown, Iowa, U.S.A., IA 50108. Tel: (515) 752 0600 |
| **GLC** | • | Greenlooms Classics, Greenlooms Farm, Hargrave, Chester, Cheshire, CH3 7RX. Tel: 01829 781636 |
| **GSO** | § | Gold Star Owners Club, c/o George Chiswell, 43 Church Lane, Kitts Green, Birmingham, West Midlands, B33 9EG. |
| **HOC** | § | Hesketh Owners Club, c/o Tom Wilson, 19 Stonnall Road, Aldridge, Walsall W59 8JX. |
| **HOLL** | *† | Holloways, 49 Parsons Street, Banbury, Oxfordshire, OX16 8PF. Tel: 01295 253197 |
| **HRM** | § | Historic Raleigh Motorcycle Club, c/o R Thomas, 22 Valley Road, Solihull, West Midlands, B92 9AD. |
| **IMC** | § | Indian Motorcycle Club, c/o John Chatterton (Membership Secretary), 183 Buxton Road, Newtown, New Mills, Stockport, Cheshire, SK12 3LA. Tel: 01663 747106 |
| **IMO** | § | Italian Motorcycle Owners Club, c/o Rosie Marston (M'ship Sec) 14 Rufford Close, Barton Seagrove, Kettering, NN15 6RF. |
| **LDM** | § | London Douglas Motorcycle Club, c/o Reg Holmes (M'ship Sec), 48 Standish Avenue, Stoke Lodge, Patchway, Bristol, Avon, BS12 6AG. |
| **LF** | * | Lambert & Foster, 77 Commercial Road, Paddock Wood, Kent, TN12 6DR. Tel: 01892 832325 |
| **MR** | *† | Martyn Rowe Truro Auction Centre, Calenick Street, Truro, Cornwall, TR1 2SG. Tel: 01872 260020 |
| **MSMP** | • | Mike Smith's Motoring Past, Chiltern House, Ashendon, Aylesbury, Bucks. Tel: 01296 651283 |
| **MVA** | § | MV Agusta Owners Club, c/o Ray Gascoine, 7 Lowes Lane, Wellisbourne, Nr Warwick, Staffordshire, CV35 9RB. |
| **MVT** | § | Military Vehicle Trust, PO Box 6, Fleet, Hampshire, GU13 9PE. |
| **NAC** | § | National Autocycle & Cyclemotor Club, c/o Rob Harknett, 1 Parkfields, Roydon, Essex, CM19 5JA. |
| **NLM** | •† | North Leicester Motorcycles, Whitehill Road, Ellistown, Leics. LE67 1EL. Tel: 01530 263381 |
| **NOC** | § | Norton Owners Club, c/o Dave Fenner, Beeches, Durley Brook Road, Durley, Southampton, Hampshire, SO32 2AR. |
| **ONS** | * | Onslows, Metrostore, Townmead Road, London, SW6 2RZ. Tel: 0171 793 0240 |
| **PC** | | Private Collection |
| **PCC** | | Cotton's Classic Bikes, Phil, Victoria Road Museum, Ulverston, Cumbria, LA12 0BY. Tel: 01229 586099 |
| **PM** | •† | Pollard's Motorcycles, The Garage, Clarence Street, Dinnington, Sheffield, Yorkshire, S31 7NA. Tel: 01909 563310 |
| **PS** | * | Palmer Snell, 65 Cheap Street, Sherbourne, Dorset, DT9 3BA. Tel: 01935 812218 |
| **PVE** | § | Preston Vintage Enthusiasts, Lancashire. |
| **REC** | § | Rudge Enthusiasts Club, c/o Colin Kirkwood, 41 Rectory Green, Beckenham, Kent, BR3 4HX. Tel: 0181 658 0494 |
| **Rod** | •† | Organ, Rod, Sporting Images, 35 The Wicket, Hythe, Southampton, Hampshire, SO45 5AU. Tel: 01703 846279 |
| **S** | *† | Sotheby's, 34-35 New Bond Street, London, W1A 2AA. Tel: 0171 493 8080 |
| **SOF** | § | Sunbeam Owners Fellowship, PO Box 7 , Market Harborough, Leicestershire, LE16. |
| **ST** | • | Steve Tonkin Restorations, North Road Garage, Lower North Road, Carnforth, Lancashire, LA5 9LJ. Tel: 01524 733222 |
| **SW** | • | Spinning Wheel Garage, Sheffield Road, Sheepbridge, Chesterfield, Derbyshire, S41 9EH. Tel: 01246 451772 |
| **TMSC** | § | Triumph Motorcycle & Scooter Club, Wales. |
| **Vel** | § | Velocette Owners Club, c/o David Allcock, 3 Beverley Drive, Trinity Fields, Stafford, ST16 1RR. |
| **VER** | •† | Verrall, Brian, Sussex. Tel: 01403 891892 |
| **VMCC** | § | Vintage Motor Cycle Club, Allen House, Wetmore Road, Burton-on-Trent, Staffordshire, DE14 1SN. Tel: 01283 540557 |
| **WL** | * | Wintertons Ltd, Lichfield Auction Centre, Wood End Lane, Fradley, Lichfield, Staffordshire, WS13 8NF. Tel: 01543 263256 |

# CONTENTS

# HOW TO USE THIS BOOK

*Miller's Classic Motorcycles Price Guide* presents an overview of the classic motorcycle marketplace during the past twelve months. In order to give you a comprehensive feel for what is available, we have included illustrations from a wide range of auction houses, dealers, motorcycle clubs and private individuals.

Following Miller's format, motorcycles are presented alphabetically by marque and chronologically within each group. Sidecars, specials, mopeds and scooters are dealt with in the same way at the end of the book. In the motorcycle memorabilia section, objects are grouped alphabetically by type, for example sign and petrol pumps, and then, where possible, chronologically within each grouping. Each illustration is fully captioned and carries a price range which reflects the dealer's or auctioneer's sale price. The prefix 'Est.' indicates the estimated price for the motorcycles which remained unsold at auction. Each illustration also carries an identification code which enables the reader to locate its source in the 'Key to Illustrations'.

We do not illustrate every classic motorcycle ever produced. Our aim is to reflect the marketplace, so if, for example, there appears to be a large number of Triumphs and only a few Vincents, this is a reflection of the quantity, availability and, to an extent, the desirability of these motorcycles in the marketplace over the last twelve months. If the motorcycle you are looking for is not featured under its alphabetical listing, do look in the colour sections and double-check the comprehensive index at the back of the book. If a particular motorcycle is not featured this year, it may well have appeared in previous editions of *Miller's Classic Motorcycle Price Guide,* which provides a growing visual reference library.

Please remember Miller's pricing policy: we provide you with a price GUIDE and not a price LIST. Our price ranges, worked out by a team of trade and auction house experts, reflect variables such as condition, location, desirability, and so on. Don't forget that if you are selling, it is possible that you will be offered less than the price range.

Lastly, we are always keen to improve the content and accuracy of our guides. If you feel that a particular make or model or other aspect of classic motorcycles has not been covered in sufficient detail, or if you have any other comments you would like to share with us about our book, please write and let us know. We value feedback from the people who use this guide to tell us how we can make it even better for them.

# RESTORATION TIPS

Not everyone can afford the professional services of a classic motorcycle restorer nor does everyone possess the skill and knowledge to take on a complicated restoration project. A combination of do-it-yourself and the employment of professional services could provide rewarding results in terms of hands-on restoration experience gained and the satisfaction of having a quality finished product.

Before you begin any restoration, photograph the motorcycle from all sides noting cable routes, wiring harness layout, the positioning of dip switches, horn button and other small details. Finishing touches are what make the difference between a good restoration and an excellent one.

An inexperienced enthusiast could tackle a restoration project armed only with a simple tool box and a workshop manual, but a parts catalogue is also vital. It will show you every last component, down to spacers and decals, so you will be aware of what is missing.

Catalogues are available from dealers who advertise in magazines such as *Classic Motorcycle*. There are also books available, dedicated to one make motorcycle renovation, obtainable from dealers who specialize in motorcycle books. They can be found at the larger auto-jumbles or classic motorcycle shows such as The Classic Motorcycle Show at Stafford in April and others throughout the year. Alternatively, visit the National Motorcycle Museum in Solihull, near Birmingham, where you will undoubtedly find a glorious example of what you are dreaming of owning.

When you come up against a problem that literature and photographs cannot help you with, such as judging whether a gearbox main shaft is worn out or not, or fault diagnosing a wiring problem, this is the time to seek professional advice. Most good restorers can be found from reputation and long existence, but the workshop charging £30 an hour will not necessarily do a better job than the workshop charging £10 an hour. Shop around and study workmanship already carried out.

The term 'specialist' means people or firms who specialize in certain makes of classic motorcycles. For instance, Stewart Engineering in Leamington Spa specialize in Sunbeams while Hughie Hancox in Coventry are specialists in Pre-unit Triumph Twins. They can give you technical advice or supply parts difficult to find. Many of the classic motorcycles clubs will know of specialist marque restorers.

Most professional workshops farm work out such as chroming, stove enamelling, wheel building or upholstery – it is unlikely they will have all these facilities under one roof. You could farm these services out yourself to save money. They should be listed in local Yellow Pages or the classified columns of classic motorcycle magazines.

Finding parts can prove to be a great problem even to the experienced restorer. Many missing parts will not be available new, even through specialists, so you will have a choice of making do with a suitable replacement by scouring the auto-jumbles or have the part specially made.

If you can find a suitable replacement it should look as similar as possible to the missing part. For instance, a cosmetic styling strip on a petrol tank might be found in black and you would have to have it chromed to make it look authentic, or a suitable replacement for a mechanical part might have to be machined to make it fit more precisely. Wading through acres of auto-jumble can be a weary task but made rewarding if you do discover that crucial left-hand footrest that is missing from your otherwise complete BSA B33. When you come across that hard to find part at an auto-jumble and the price is not insanely high – buy it! I have known of a 1972 BSA A65 Firebird Scrambler to be comprised entirely from auto-jumble parts and a gentleman who searched for three years for a Triumph T100 parking light lens only to find it at an auto-jumble. Publications such as *Classic Bike* or *British Bike* will list auto-jumble events across the whole country.

If there is no other alternative than to have the missing part specially made there may be an engineering company locally who can provide this service but they will need a pattern or template, in which case you will have to borrow a donor part. Engineering and machining labour is quite expensive but you may not have a choice.

As missing parts can often be difficult to locate and expensive to make, try to find a more complete example to restore. Many motorcycles 'turn up' original, in one piece and absolutely complete. A dealer would advertise this as a 'very good restoration project' and ask a higher price for it but in the long run it will be so much easier to restore, particularly if it is your first project.

Any machine can be restored as long as it is restorable! Even a hybrid or a special, such as a Triton, could be obtained as a wreck and brought back to its former glory, but if the machine has sat in a pond for years the 'tin wear' could be completely unrestorable and perhaps only serve as a pattern to have a new part made.

In short, if you are confident and have time to do much of the restoration, have a go yourself. If you are not and you can afford it, leave it to the experts!

John Newson

# THE MOTORCYCLE MARKET

The motorcycle collector's market has traditionally been for enthusiast's, far less subject to the whims and excesses of fashion and the ups and downs of financial markets, and unlike the collector's cars scene, never invaded to any great degree by 'investors'. Because of this, motorcycle enthusiasts tend to be more 'hands on' and knowledgeable than their car counterparts, more particular about originality, better able to detect when a bike is not correct, and unwilling to pay top money for any machine which is less than top quality.

Thus, motorcycles were not in general subject to the spiralling prices of the late eighties and the plunging markets of the early nineties, although machines in the top quality class like Brough Superior and Vincent were affected to a degree. It would be true to say that the early nineties have witnessed a steady and healthy market with no shortage of enthusiasm, and since the 'single market' became a reality there has been continued growth with an increasing number of Continental buyers both participating in and attending UK motorcycle sales – often prepared to bid higher figures than their UK counterparts to obtain the machine of their choice.

Very specialised machines, however, continue to have a very limited market, and this is particularly true of Japanese racing motorcycles. Even those with impeccable provenance in the shape of letters from manufacturers, team managers and riders, are failing to find buyers. If there is one area, however, which never fails to inspire enthusiasts, it is the 'barn discovery'. The lure of Tutankhamen's tomb is as strong today as it was in 1922 and because motorcycles are smaller than cars and can be hidden away in quite small spaces, they continue to emerge in encouragingly large numbers.

The manner in which motorcycles are marketed may also be seen to have a very definite influence on the prices they achieve. In this respect it is significant that a popular venue, a good catalogue, an international mailing list, a good selection of quality machines and a competent auctioneer or dealer, with good presentation, can make the difference between good prices and disappointment.

Recent sales of motorcycles since last December have graphically underlined all these points. At Sotheby's at the Royal Air Force Museum in December, a wide variety of good quality machines, both restored and unrestored, was offered, with excellent results. A 1973 MV Agusta 750S with just 12,000 miles on the clock and in commensurate condition brought £14,000, whilst a 'barn discovery' 1927 Brough Superior 680cc machine realised £8,400, and a 1955 Vincent Black Prince attracted an Austrian bid of £12,200. The sale was also significant in that it heralded an increasing interest from Japan, not just on the 'superbikes' of the British industry, but on early Japanese machines and good quality British touring machines, and cycle motors.

Whilst Germany, Holland, Spain, Greece and, to an extent, the USA, have all been active, the Japanese have often proved to be the strongest bidders, and their interest in cycle motors (those little bolt-on engines with which people motorised their bicycles in the 1940s and 1950s) has been increasing. At Brooks sale at the National Motor Cycle Museum in July, all the cycle motors were strongly bid, but at the other end of the sale a 1912 Rover Combination with a sidecar in wickerwork brought £6,000, a beautifully restored 1914 Indian board track racing machine made £8,280, and a low mileage example of a 1981 Triumph T140 Royal Wedding (a limited commemorative edition) Bonneville and a superb Trident made £3,910 and £5,175 respectively. Barn discoveries included a 1927 AJS V-twin, off the road since 1948 and with one family from new, which realised £2,850 despite needing full restoration.

If any trends may be detected from these recent sales, it is that there has been some falling off in the numbers of UK private buyers bidding at sales and attending them. Their place has been taken, to a very large degree, by the increased number of overseas buyers and reports from the trade indicate that dealers are having the same experience.

It would seem that what Chancellor Kenneth Clark perceives as the 'missing feel good factor' may, at last, be beginning to touch the UK motorcycle enthusiast, but it is difficult to reconcile this hypothesis with the very high prices still being paid for top quality machines, both restored and unrestored. What is clear, however, is that the market has a surfeit of mediocre machines for sale largely as a result of the early 1990s' recession. Japanese racing machinery is currently overpriced. With their traditionally shrewd eye for a bargain, British bikers are obviously saving their money for the better quality machines, more of which are now finding their way to the UK because of the single market. This is particularly true of Italian machines, and a 1949 MV Agusta Competizione imported for the Brooks sale sold to Japan for £6,000, whilst Sotheby's sold a fine Ducati Mike Hailwood replica last December for £4,300.

Brooks and Sotheby's now dominate the motorcycle auction market, largely eschewed by the other auction houses, and have already clocked up sales of over half a million pounds this year between them. With other sales scheduled, the future looks promising.

Malcolm Barber
Managing Director – Brooks Auctioneers Ltd

# ABC *(British)*

**1922 ABC Sopwith 398cc,** transverse-mounted horizontally opposed twin, with overhead valves, steel cylinders, cast-iron hemispherical heads in unit with a car-type gearbox, the frame features suspension by quarter-elliptic springs.
**£1,400–1,600** *S*

**1921 ABC 3hp 398cc,** leaf-sprung frame, car-type 4-speed gearbox and 2 wheel hub brakes, original log book and a 1930 licence disc.
**£4,000–4,500** *BKS*

*Production problems meant that the ABC never fulfilled its early promise. Early examples had a tendency to shed pushrods, and manufacture ceased in 1923 after only 3,200 machines had been built.*

# ACME *(British)*

**1913 ACME Solo,** 3½hp side valve, Precision engine with direct belt drive, restored to present condition in 1965, finished in black with white petrol tank and red detailing, period horn fitted, in good mechanical condition.
**Est. £4,000–5,000** *S*

*The Coventry based Acme Company, founded in 1902, amalgamated with the Rex Company in 1922 to form Rex-Acme. In the years prior to this they used a variety of proprietary engines in their products.*

# ADLER *(German)*

**1958 Adler Model MB250S 247cc,** twin cylinder 2-stroke.
**£1,700–2,000** *C(A)*

*This machine was one of the very last to be built by this famous German manufacturer. The MB250S was the top-of-the-range sports model and differed from the earlier M250 and MB250 by way of its swinging arm rear suspension, in place of the original plunger type. The motor delivered 18bhp with a top speed of 77mph. The Adler is seen today as the forerunner of the modern roadster 2-stroke twin and Yamaha used the MB250S as a basis for its successful air-cooled twins of the 1960s and*

**1956 Adler Model MB 250cc,** twin cylinder engine and 4-speed gearbox, one of only 3 known surviving air-cooled road going Adlers in the UK, fair condition, paintwork poor, frame and tyres good, non-original seat and exhaust system, together with *Motor Cycling* road tests, owner's manual, history and V5.
**£400–650** *S*

*Better known these days for their typewriters and office equipment, Adler of Frankfurt built both cars and motorcycles, the latter from 1900 until 1957. The post-war range of 2-stroke singles and twins included machines from 98cc up to 250cc fitted with their own engines, and included some very successful 250cc racing versions, many of which had water-cooling.*

# AERMACCHI
## (Italian 1950–1972, Harley-Davidson 1960–1978

Although it never won a world title under its own name, Aermacchi is still a famous name in Italian motorcycle racing history. In addition, under the Harley-Davidson label, Aermacchi 2-strokes gained four titles in the mid-1970s, with Walter Villa as rider.

The company, based in Varese, had a long history, but much of it in the aviation arena. It was not until 1950 that Aermacchi entered the 2-wheel world.

For the first decade it concentrated its efforts on roadsters, albeit with the odd dirt bike or record breaker thrown in. Then in 1960 two important things happened. The first was a financial tie-up with the American Harley-Davidson concern; the second when teenager Alberto Pagani finished 9th in the Dutch TT on what was essentially a converted roadster. A week later in Belgium he did even better with a 5th. Spurred on by this success,

Aermacchi decided to build 'over-the-counter' versions for sale to private customers.

Although the various capacities of this humble pushrod single never won a Grand Prix, they nonetheless formed the backbone of privateer racing during the mid-1960s, between the demise of the British singles and the onslaught of the Yamaha TD, TR and later, TZ twins.

Realizing the limitations of the design, the Varese marque then moved into 2-strokes. First came a 125 single in 1968, followed two years later by the prototype of the twin cylinder model that was to prove so successful in later years.

These developments were mirrored in the roadsters: the Ala Verde (Sprint in the USA) flat single 4-stroke and later SS350, making way for a variety of 2-strokes in the mid- to late 1970s.

**1960 Aermacchi Harley-Davidson Ala Verde 250cc,** OHV horizontal single, restored, condition concours.
**£2,500–3,000** *BCB*

**1962 Aermacchi Ala d'Oro Racer 246cc,** single cylinder, 4-stroke.
**£6,000–7,000** *C(A)*

*The American Harley-Davidson Company gained a 50% stake in Aermacchi during 1960. This resulted in a significant increase in production at the company's Varese works. In addition, Aermacchi built a series of motorcycles including this 1962 example of its famous Ala d'Oro model. Up to the end of 1963 the engine was of long-stroke design with a bore and stroke of 66 x 72mm; thereafter the dimensions were changed to 72 x 61mm, giving a new capacity of 248cc.*

**1962 Aermacchi Ala Verde 246cc,** OHV 4-speed.
**£2,500–2,800** *PC*

**1964 Aermacchi 250cc.**
**£4,000–4,500** *VER*

*l.* **1971 Aermacchi TV 344cc,** Italian import.
**£2,000–2,200** *PC*

**1968 Aermacchi Ala d'Oro 344cc Racer,** OHV, 5-speed, Ceriani suspension. **£6,500–7,000** *AtMC*

**1974 Aermacchi Model HD-SS 350 344cc.** **£1,100–1,500** *BKS*

**Use the Index!**

*Because certain items might fit easily into any number of categories, the quickest and surest method of locating any entry is by reference to the index at the back of the book.*

*This index has been fully cross-referenced for absolute simplicity.*

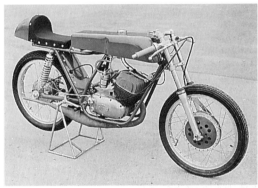

**1970 Aermacchi 125 Racer,** 5-speed, 28mm carburettor, 110 mph. **£3,500–3,800** *PC*

**1972 Aermacchi TV 344cc,** 5-speed. **£2,200–2,500** *PC*

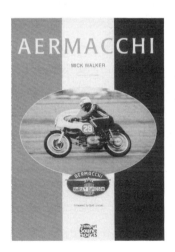

**1974 Aermacchi RR250,** 2 stroke twin cylinder, replica of factory GP bike. **£4,000–4,200** *PC*

# AJS *(British 1909–1966, FB AJS from mid-1970s)*

The Stevens brothers, Harry, George, Jack and Joe, began to experiment with internal combustion engines in 1897 at their father's Wolverhampton based engineering works. They put this to good use by initially building engines which were used in some of the earliest motorcycles, cyclecars and even cars. Then in 1909, using Jack's initials (Albert John Stevens), the four brothers formed the AJS Company.

Having won their first TT, the Junior, in 1914 the firm returned to the racing scene after the Great War to become outstandingly successful in the 350cc class both at home and in Europe. The marque won three Junior TTs in succession, in 1920, 21 and 22, whilst in 1921 Howard Davies, riding a 350 OHV AJS, became the first rider to win a Senior TT on a Junior machine, a feat which has never been repeated.

With racing becoming ever more competitive, AJS developed a new overhead cam works model for 1927. But somehow even though the bike was reliable, it simply wasn't quick enough to consistently beat the best of the Rudge, Sunbeam and Norton machinery of the era.

Like so many others, AJS was also deeply affected by the worldwide economic depression. By mid-1931 the Stevens' empire was on the rocks and the factory doors closed. This was not to be the end for AJS and the company was acquired by the Collier brothers, already owners of the south-east London based Matchless marque. This new group was to become known as AMC (Associated Motor Cycles), who subsequently swallowed up Sunbeam, James, Francis-

**1924 AJS Big Port 350cc.**
**£4,800–5,200** *BLM*

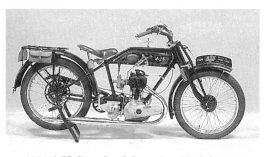

**1924 AJS Standard Sporting Model 5 349cc,**
single cylinder, 4 stroke, SV.
**£2,500–2,800** *C(A)*

*An AJS model Sporting 5 with IOE (inlet over exhaust) engine. Models such as this established the Wolverhampton marque's reputation as a builder of high class single cylinder motorcycles for both road and track.*

**1925 AJS OHV Big Port 358cc.**
**£6,500–6,800** *VER*

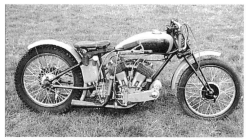

**c1928 AJS 1000cc,** V-twin engine.
**Est. £4,000–5,000** *BKS*

*l.* **1930 AJS R4 350 SV,** restored, dating certificate issued by the AJS and Matchless Owners' Club.
**Est. £2,000–2,250** *S*

*Produced in 1930 as a 350cc single side value engined middleweight it was intended to fill the market as a deluxe sporting version of the range for that year.*

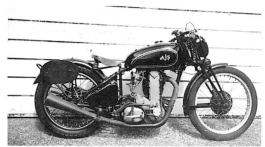

**1930s AJS R7 OHC Racer 350cc.**
**£6,000–6,500** *BLM*

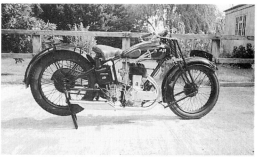

**1931 AJS Model 9 500cc.**
**£2,700–2,900** *BLM*

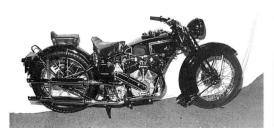

**1933 AJS V-Twin 990cc,** Big 2 (33/2),
fully restored.
**£5,000–6,000** *PC*

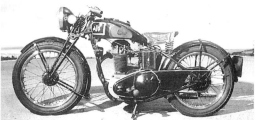

**1935 AJS Model 26 350cc,** pre-war single cylinder
overhead valve 4-stroke engine, 4-speed gearbox,
girder forks and rigid frame at rear, no known
modifications from the original maker's
specification, spare engine, good condition.
**£1,700–2,000** *S*

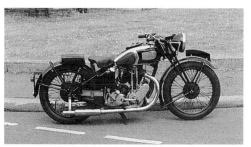

**1937 AJS Twin Port 350cc.**
**£2,300–2,700** *PM*

*This 1937 350cc Twin Port features the authentic
chrome plated tank which reflects the improved
times as the decade progressed. The timing chain
case is an early example of the design used on
later models including the famous G3/L. All
AJS models by this year, except the export Vee-
twin, featured footchange gearboxes and most
employed magneto ignition.*

**1948 AJS Model 16M 350cc.**
**£1,800–2,400** *BLM*

**1951 AJS 18S 498cc,** mileage just over 16,000,
good condition overall.
**£2,300–2,800** *BKS*

*The 500cc Model 18 was announced soon after the
end of the war, and is a typical British big single.
It features a vertically divided crankcase, built-up
crankshaft and right-hand timing gear, tall
pushrod tunnels and a chain driven magneto
driven by the chain from the exhaust camshaft and
mounted ahead of the engine.*

**1951 AJS Model 16M 350cc**
**£2,000–2,300** *BLM*

**1953 AJS 7R 350cc Racing**
**Est. £15,000–16,000** *BKS*

**1953 AJS Model 20 Twin 500cc,**
fully restored.
**£1,700–2,000** *MR*

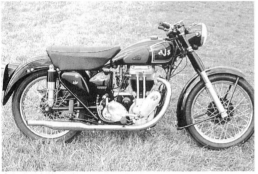

**1953 AJS Model 18S 500cc,** extensively
restored to concours condition, original
registration number.
**Est. £2,800–3,000** *BKS*

**1955 AJS 18S 500cc.**
**£1,500–2,000** *PS*

**1957 AJS 7R 350cc.**
**£12,500–13,000** *VER*

**1958 AJS Model 18S 498cc,** single cylinder,
OHV, 4-stroke, 21bhp.
**£2,400–2,700** *C(A)*

*From sporting origins in the 1920s, the Model
18 was one of the longest running models in the
AJS range. Rugged road reliability and ease of
maintenance were its selling points.*

**1958 AJS 16MS 350cc,** very good condition, V5,
valid MOT.
**£1,750–2,000** *BKS*

*The Plumstead factory under the title of AMC
produced many 'badge' engineered machines and
attempted to keep models with their own identity.
The result was quite successful with each brand
name generally having its own following.*

**1960 AJS Model 8 348cc,** finished in blue and
black, to catalogue specification.
**Est. £1,200–1,600** *S*

*The 348cc Model 8 offered a similar level of
performance to the sporting 250 CSRs in a less
highly stressed package, and featured heavier
Teledraulic front forks.*

**1961 AJS Model 8 350cc.**
**£1,000–1,400** *BLM*

**1960 AJS Model 16 350cc,**
finished in AJS black livery, with
dual seat, sprung frame and full
lighting set, full restoration to
concours standard, displayed in
a private collection.
**£2,000–2,500** *BKS*

**1963 AJS Model 14 CSR 248cc,**
wheels, mudguards, dual seat, tank
and engine all period.
**£650–850** *ADT*

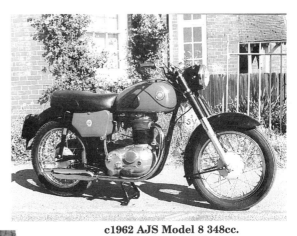

**c1962 AJS Model 8 348cc.**
**£1,300–1,800** *AtMC*

*The AJS (and Matchless) Lightweight*
*350 first appeared in 1960 and ran*
*through to the end of 1962.*

**1958 AJS Model 16C Trials,** fully restored to
original specification, including 'jam pot' rear
shocks and upswept silencer.
**Est. £2,300–2,500** *PC*

**c1970s AJS Stormer Motocross**
**Scrambler 370cc.**
**£800–1,100** *BLM*

# AMBASSADOR *(British)*

**1952 Ambassador 197cc.**
**£350–600** *WL*

*This 1952 Series V Ambassador with 197cc*
*Villiers 6E engine and 3-speed gearbox is*
*typical of early post-war British 2-strokes.*

# ARIEL *(British 1902–1970)*

In 1902 Ariel began to manufacture motorcycles although they started building three-wheelers in 1898. Beginning first with a 3.5hp single cylinder, by WWI the Ariel range had expanded to include 498cc side valve singles and IOE (inlet over exhaust) V-twins.

Ariel, then owned by Jack Sangster, was joined in 1927 by the young engineer Edward Turner. Chief Designer Val Page was also an important part of Ariel's success designing new 498cc OHV and 557cc OV singles.

The famous Square Four, at first only in 498cc form, was introduced in 1931. Designed by Turner, larger Square Fours with 601 and 996cc engines followed later in the decade. In 1936 Edward Turner transferred to Triumph in Coventry, after Jack Sangster bought that factory. During the war, many 347cc OHV Ariels were used by the British army.

Post-war production centred on 347 and 497cc OHV Red Hunter singles, the Huntmaster vertical twin and a revised 997cc OHV Square Four. Other models of the early 1950s, Ariel's range included the BSA derived 200 Colt single and a 598cc SV single intended mainly for sidecar use.

Jack Sangster who had also been involved with BSA since the late 1940s, moved Ariel into the BSA group. This meant that several Ariel models sported BSA parts during this time. Eventually Ariel was transferred to BSA's Selly Oak, Birmingham, factory.

There the famous 247cc vertical twin 2-strokes, the Leader and the Arrow, were designed. However, the 197cc version was to be last Ariel model (if the Ariel 3 moped is discounted) to be built and the marque ceased to exist at the end of the 1960s.

**1920 Ariel 7hp V-twin,** belt drive, complete, requires restoration, old and new documents.
**£4,200–4,800** *S*

**1928 Ariel Model A 500cc.**
**£3,000–3,300** *BLM*

**1936 Ariel Red Hunter 250cc,**
restored, with no known deviations from catalogue specification, Swansea V5 and MOT.
**Est. £3,500–4,000** *S*

**1929 Ariel Brooklands 500cc.**
**£3,800–4,300** *PM*

*A rare Brooklands racing model, this 500 Ariel single sports OHC with magneto ignition was soon snapped up when it went on sale, reflecting its unique nature.*

Miller's is a price GUIDE
not a price LIST

*l.* **1936 Ariel Red Hunter 250cc,**
excellent condition.
**£2,500–3,500** *SW*

**1939 Ariel Red Hunter OHV Single Cylinder 346cc,** good condition, new tyres and wheels, the petrol tank resprayed, now fitted with a 1953 engine and non-standard in some respects.
**£650–750**  *PS*

**1939 Ariel VH Red Hunter 497cc,** OHV single cylinder, twin port with separate high level exhaust, chrome plated tank, 4-stroke, 24.6bhp, Webb girder forks.
**£4,000–4,500**  *C(A)*

*Stylish good looks and sparkling performance made the Ariel Red Hunter one of the most sought after motorcyles of the late 1930s. This is one of the last to be produced before war broke out in September 1939.*

**1939 Ariel NH Red Hunter 350cc.**
**£1,500–2,500**  *BLM*

**1940 Ariel W-NG 350cc.**
**£2,200–2,300**  *BLM*
*Popular girder War Department bike.*

**1947 Ariel SV 600cc,** restored.
**£2,000–2,500**  *SW*

**1948 Ariel Red Hunter 350cc.**
**£1,800–2,400**  *BLM*
*An all-round club run single.*

**1948 Ariel Square Four,** reasonable condition requiring some attention.
**Est. £2,000–2,500**  *S*

*Nicknamed the 'Squariel', the Ariel Square Four was the work of Edward Turner, which was first made with a 498cc engine in 1931. The following year it was joined by a 601cc version achieved by boring out the cylinders from 51 to 56mm, the stroke remaining unchanged at 61mm, with the introduction of an even larger model (the 995cc) 4G in 1937.*

**1948 Ariel Square Four 1000cc.**
**£3,000–3,500**  *BLM*
*Equally suitable for solo or sidecar use.*

**1949 Ariel VH Red Hunter 500cc.**
**£2,400–2,600**  *BLM*

**1951 Ariel Single Cylinder 500cc,** good
condition throughout and finished in red and
black, sprung saddle and pillion pad, Anstey link
rear suspension and telescopic front forks.
**£1,400–1,600**  *S*

**1952 Ariel Square Four 1000cc,** Two Pipe
model, finished in maroon/black, Solex
carburettors, oil cooler and coil ignition,
very good condition throughout.
**£2,000–2,300**  *MR*

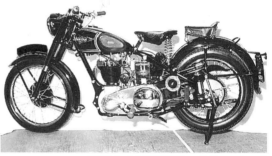

**1952 Ariel NH 350cc,** fully restored.
**£2,000–2,500**  *AOC*

**1954 Ariel Square Four 997cc,** 4 cylinder,
OHV 4 stroke.
**£6,000–6,500**  *C(A)*

*Known as the 'Squariel', the 997cc 4G was not
only used for solo use, but was also a popular
choice with sidecar men. Ariel Square Fours
were in production from 1931–59 in an
assortment of engine sizes. This is the definitive
4G MK2, built from 1953–59.*

**1952 Ariel VB 600cc.**
**£2,400–2,600**  *BLM*

**1955 Ariel Huntmaster 650cc,** restored using
original parts.
**£1,800–2,000**  *BKS*

*The Huntmaster was the name given to Ariel's
largest capacity twin, its bigger sister was the
Square Four.*

**1955 Ariel Red Hunter Model NH 350cc,**
single cylinder OHV 4-stroke engine, Burman
4-speed posi-stop gearbox, telescopic forks and
sprung frame at rear, good unrestored
condition throughout, original tax disc and old
style buff log book.
**£1,200–1,500**  *S*

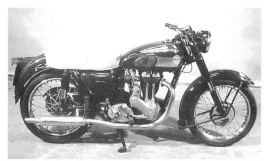

**1955 Ariel NH 350cc,** V5 and
old MOT certificates.
**Est. £1,850–2,000** *S*

**1957 Ariel KH Fieldmaster 500cc.**
**£2,000–3,000** *BLM*

**1958 Ariel Square Four.**
**Est. £3,800–4,200** *BKS*

**1958 Ariel Leader Solo,** finished in aqua and
light grey with many Ariel extras, including
8-day clock and luggage straps, restored in 1980,
work included rebuilt crank, bearings, oil seals,
new chain, clutch, alternator, wiring loom,
electrical parts, tyres, screen, exhaust and
silencers fitted, factory tools.
**Est. £700–1,100** *ALC*

*This is an exceptional example of these popular
vertical twins, and has an outstanding history. It
is believed to be the only surviving 1958 Leader.*

**1959 Ariel Arrow 250cc.**
**Est. £300–400** *S*

*Whilst the Leader met with considerable
success, there were those who did not want the
full enclosure but appreciated the machine's
excellent handling and performance. The
Arrow was born featuring a dummy fuel tank
and a slightly tuned motor.*

**1961 Ariel Golden Arrow 250cc,**
converted to Café Racer trim.
**£500–600** *MR*

**1960 Ariel Leader 250cc.**
**£1,200–1,500** *BLM*

*r.* **1962 Ariel Golden Arrow 250cc,**
fully restored.
**£1,500–1,600** *AOC*

**1962 Ariel Arrow 2 Stroke Twin Cylinder 249cc.**
£850–950 *PS*

**1964 Ariel Arrow 200cc.**
£800–1,500 *BLM*
*One of the rare smaller Arrows – worth collecting.*

Miller's is a price GUIDE not a price LIST

*l.* **1965 Ariel Super Sports Arrow 250cc.**
£1,500–2,000 *BLM*
*This model is often known as the Golden Arrow.*

# BIANCHI *(Italian)*

**1929 Bianchi Sport 250cc.**
£1,600–2,000 *PM*

# BIMOTA *(Italian 1972–)*

Bimota entered the motorcycle business by making frame kits for Honda and MV 4 cylinder engines in the early 1970s. By 1976 the firm was producing complete machines, usually with Suzuki engines. Thereafter came associations with Kawasaki, Honda, Yamaha and finally Ducati.

**1976 Bimota (Harley-Davidson) RR250 Racing.**
£4,800–5,200 *BKS*

# AUTOMOTO *(French)*

**c1928 Automoto 250cc,** powered by a Zurcher of Paris 248cc single cylinder, air-cooled, OHV engine, hand change 2-speed gearbox, auxiliary oil hand pump and sports handlebars, original black livery and tool box, luggage carrier and girder fork and coil spring front suspension.
**Est. £800–1,000** *BKS*

**c1976 Bimota (Harley-Davidson) 500cc Racing,** 90hp, 4 carburettor, water-cooled, twin cylinder engine, built by Aermacchi/Harley-Davidson, anti-dive front brake set-up, twin discs geared to rotate in the opposite direction to the wheel.
£14,000–14,500 *BKS*

# BMW *(German 1920–)*

Although BMW (Bayerische Motoren Werke) can trace its history back to the 19thC, the first BMW motorcycle did not make an appearance until after the end of the WWI. Strictly speaking this began in 1920 with development work on a machine using a proprietary Kurier 148cc 2-stroke engine. Named the Flink, it was not a commercial success. Next, in 1921, came the introduction of an engine which was to shape BMW's two-wheel future. Designed by Martin Stolle it was a 493cc flat twin SV 4-stroke. Besides being used by BMW to power its own Helios machine it was also supplied to rival firms such as Victoria and Bison.

The first real BMW came in 1923, when the R32 was launched at the Paris *Salon*, powered by a direct descendant of the 1922 MZ B15 engine from the Helios. A major difference was it being mounted transversely in unit with a 3-speed gearbox and shaft final drive (the Helios featured not only fore-and-aft mounting, but chain drive).

The R32 was the beginning of a virtually unbroken line which still exists today, albeit in modernised form. The only other engine layouts being various vertical singles which spanned the late 1920s through to the late 1960s; and the K-series 3 and 4 cylinder models which began with the K100 of 1983.

From 1928–1950 BMW built supercharged racers which were used by a number of riders including Georg Meier and Walter Zeller. When Germany became a member of the FIM (Fédération Internationale Motocycliste) in 1951 and unable to use superchargers a special version of the flat twin engine with overhead cams instead of overhead valves, found its niche in sidecar racing; going on to win no less than 19 championships. Drivers included Max Duebel, Florian Camathias, Fritz Scheidegger and Fritz Hildebrand.

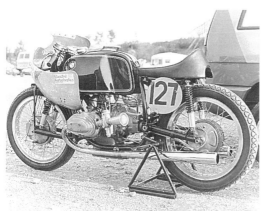

**1954 BMW Rennsport Racer 494cc,** DOHC, Dell'Orto carburettor and Earles forks, very rare.
**£20,000+**  *PC*

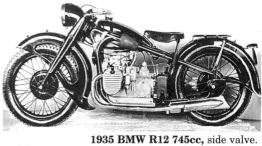

**1935 BMW R12 745cc,** side valve.
**£6,000–6,500**  *PC*
*This model was built from 1935–38.*

**1970 BMW R75/S 750cc,** 2 cylinder bore and stroke 82 x 70.6mm, 50bhp at 6500rpm, fully restored.
**£2,000–2,500**  *BMW*

*This motorcycle was first registered in Paris in 1970 and is one of 38,387 built between 1969 and 1973.*

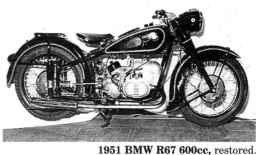

**1951 BMW R67 600cc,** restored.
**£3,000–3,500**  *BMW*

*r.* **1970 BMW R75/S 745cc,** OHV, horizontal twin, first year of this model, concours condition.
**£3,500–4,000**  *PC*

**1978 BMW R75/6 750cc,** in excellent
original condition, MOT.
**£1,000–1,300**   *MR*

*l.* **1973 BMW R90S 898cc,** OHV
twin, 5-speed gearbox.
**£3,500–4,000**   *PC*

# BRADBURY *(British)*

**c1916 Bradbury,** V-twin.
**£5,000–6,000**   *BLM*

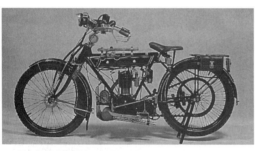

# BOWN-VILLIERS *(British)*

**1921 Bown-Villiers 269cc.**
**£2,000–2,500**   *BLM*

*l.* **1920 Bradbury 349cc,** single cylinder, side
valve, 4-stroke, bore and stroke 74.5 x 80mm.
**£5,000–6,000**   *C(A)*

*The Bradbury company produced motorcycles
at their Oldham factory from 1901–25. This
example is the last of the 2¾hp light tourers
left in existence.*

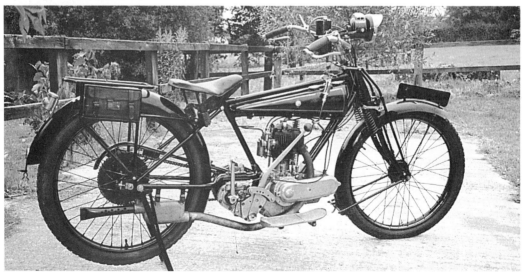

**c1921 Bradbury 2¾hp 350cc.**
**£2,600–3,000**   *BLM*

# BROUGH–SUPERIOR
*(British)*

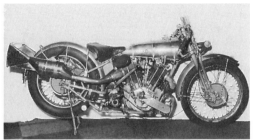

**1925 Brough Superior SS100 998cc,**
original condition.
**£35,000–40,000**  *VMCC*

*This model was used by G. W. Guyler to attain a 100mph Brooklands Gold Star in 1925.*

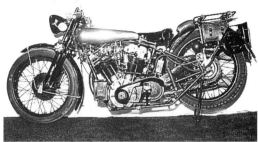

**1925 Brough Superior SS100 980cc,** restored, not original finish.
**£30,000–35,000**  *VMCC*

*This ex-Joe Wright racer is now used on the road and in 1994 completed the Austrian Alpine Rally.*

**1927 Brough Superior SS100 997cc,**
JAP V-twin engine.
**£32,000–37,000**  *AtMC*

**1929 Brough Superior SS100 Pendine 998cc,**
V-twin, OHV, 4-stroke JAP, bore and stroke 80 x 99mm.
**£35,000–40,000**  *C(A)*

*The top-of-the-line SS100 was available in various forms, but only 20 of these racing Pendines were made. A supercharged version was timed in 1936 at 169.79 mph. Lawrence of Arabia was the proud owner of six Brough Superiors.*

**1930 Brough Superior 680cc,** OHV.
**£13,500–14,500**  *VER*

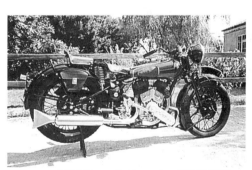

**1936 Brough Superior SS80 1000cc.**
**£14,500–15,000**  *BLM*

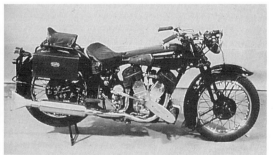

**1936 Brough Superior 1150 Solo,** side valve JAP V-twin engine, 4-speed foot change gearbox, with full electric lighting set, Brooklands fishtail exhausts, fitted toolbox and pillion seat, black paintwork, in very good condition, roadworthy.
**£9,500–10,500**  *S*

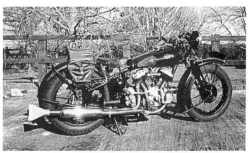

**1939 Brough Superior 1150cc,**
JAP springer, sprung frame model.
**£12,000–13,000**  *BLM*

# BSA *(British 1906–71, Late 1970s–)*

Birmingham Small Arms (BSA) began life in 1906 as a supplier of cycle parts to British and foreign factories. It was not long afterwards that they began to produce motorised bicycles and then motorcycles. By 1921 their first SV V-twin 770cc was built and successfully marketed.

Although BSA had flirted with racing by way of Brooklands and the Isle of Man TT, its real sporting successes were to come in major trials, six day events and early scrambles. In all three of these disciplines BSA made an impact and to further show off the excellence of its standard product, it became involved with demonstration tests including the famous Maudes Trophy.

So BSA grew into a large enterprise in which for so long it was able to claim 'one in four is a BSA', prospering with machines that suited showroom sales. Machines such as the Round Tank and the Sloper, two which became classics, were backed up by a series of singles and V-twins, in a variety of engine capacities.

One machine, or at least a name, gave BSA a more sporting guise. This was the Gold Star. Few motorcycles can be everything to everyone but the 'Goldie' came closer than most – a 100mph lap at Brooklands, BSA's single could cope with almost anything. Production of this bike finally ended in 1962.

After WWII the BSA group also comprised Ariel, Sunbeam and Triumph, making it the most powerful force in the British motorcycle industry until the late 1960s and early 1970s. Even so BSA built a truly vast array of models in the post war era; its most well-known being the Bantam, A-series twins in both pre-unit and unit guises, C15/B40 singles and the Rocket 3 triple. The company was reconstructed in the late 1970s to produce lightweights for export and replica Manx Norton frames.

**c1916 BSA 4¼hp,** finished in traditional green, cream and red livery with black mudguards and other cycle parts, in sound condition, buff log book.
**Est. £3,800–4,200** *S*

**1922 BSA Tourer 557cc,** full alloy chaincase, fitted with full acetylene lighting kit, period horn and leather fronted toolboxes mounted on the rear carrier, excellent condition.
**£3,000–3,500** *S*

*BSA's side valve singles offered a combination of good performance, high build quality and reasonable cost, that helped to place the company at the top of this country's motorcycle producers.*

**1925 BSA Solo 493cc,** good mechanical condition, very good black and green paintwork and fitted with full acetylene lighting set.
**Est. £2,500–2,800** *S*

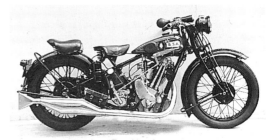

**1927 BSA Sloper 500cc,** OHV.
**£4,500–5,000** *VER*

*BSA stands for 'Birmingham Small Arms' and the products from Small Heath were soundly engineered and reliable. From 1927 the factory produced the still famous 'Sloper' model.*

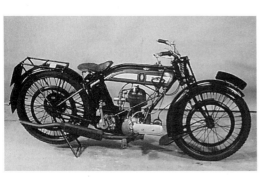

*l.* **1927 BSA Side Valve 493cc,** sound condition, black cycle parts and traditional green petrol tank with white panels lined in red and yellow.
**Est. £2,500–2,800** *S*

**1930s BSA Empire Star 250cc.**
£1,500–1,700 *BLM*

**1934 BSA Sports Star 350cc.**
£3,200–3,300 *BLM*

**1934 BSA Red Star 348cc,** single cylinder, OHV,
4-stroke, bore and stroke 71 x 88mm, 23bhp.
£1,800–2,300 *C(A)*

**1934 BSA B1 250cc.**
£2,000–2,300 *AT*

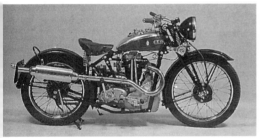

**1936 BSA Blue Star 499cc,** single cylinder, OHV,
4-stroke, bore and stroke 85 x 88mm, 23bhp.
£3,000–3,500 *C(A)*

*This model was the largest version of the long
running Blue Star range and continued the BSA
tradition of building motorcycles of simple and
inexpensive design.*

**1937 BSA G14 985cc,** V-twin, side valve,
4-stroke, bore and stroke 80 x 98mm.
£3,000–3,500 *C(A)*

*A simple and reliable workhorse built to a price.
They were used extensively as commercial
delivery vehicles and by several police forces,
including Australia.*

**1937 BSA B24 Empire Star 350cc,** restored to
concours condition, mostly original, 90,000 miles.
£3,800–4,400 *PC*

*Only seven were registered in the UK in 1991.*

**1937 BSA B26 350cc,** to original
specification in all major respects, old style
buff log book.
£1,400–1,600 *S*

*This motorcycle was first registered in
April 1937.*

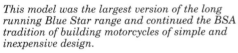

*l.* **1939 BSA M20 500cc.**
£1,600–2,000 *BLM*

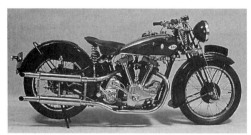

**1939 BSA Y13 748cc,** V-twin, OHV, 4-stroke, bore and stroke 71 x 94.5mm.
**£8,500–9,200**  *C(A)*

**c1939–45 BSA M20.**
**£1,200–2,000**  *MVT*

**1944 BSA WD M20 500cc.**
**£1,500–1,800**  *AT*
*This model was a workhorse of WWII.*

**1946 BSA Solo 249cc,** single cylinder, side valve, 4-stroke engine, telescopic forks and rigid rear frame, luggage carrier and lighting set, sound and largely original.
**£650–800**  *S*

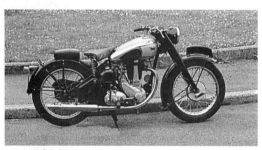

**1947 BSA ZB31 348cc,** OHV single, rigid frame, telescopic forks.
**£1,200–1,700**  *PM*

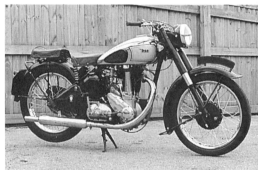

**1948 BSA B31 348cc,** totally restored.
**£2,000–2,750**  *SW*

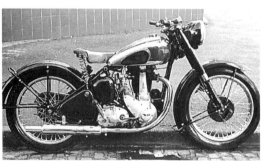

**1948 BSA B31 348cc,** OHV, single, re-built.
**£1,400–1,650**  *BCB*

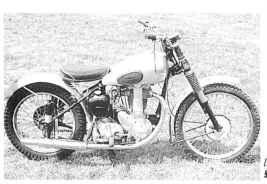

**1949 BSA A7 497cc,** original unrestored condition, except silencer.
**£1,500–2,000**  *BCB*

*l.* **1949 BSA Gold Star Trials 350cc.**
**£2,000–2,200**  *BKS*

**1949 BSA Bantam D1 125cc.**
**Est. £550–660** *S*

*The GPO bought hundreds of Bantams*
*for their telegraph boys.*

**1950 BSA A7 Star Twin 500cc,**
twin carburettor model.
**£4,400–4,600** *BLM*

**1950 BSA B31 350cc.**
**£1,700–2,300** *BLM*

**1951 BSA B33 500cc,** OHV.
**£1,200–1,500** *AT*

**1951 BSA D1 Bantam Trials 125cc.**
**£900–1,200** *BLM*

**1951 BSA B34 Gold Star 500cc.**
**£4,000–4,500** *BLM*

**1951 BSA D1 Bantam 125cc,**
unrestored original condition.
**£400–650** *BCB*

**c1952 BSA Gold Star Special 350cc.**
**£4,000–4,200** *BLM*

*This model features a McCandless*
*converted frame.*

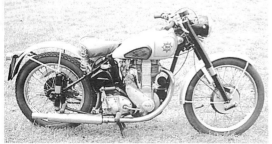

*l.* **1952 BSA Gold Star 350cc,** in Clubman
trim, mudguards, stays, chain cases,
chainguard, brake back plates and petrol
tanks painted matt silver.
**£2,800–3,200** *BKS*

*This Gold Star is one of a batch of 50 built to*
*qualify for entry into the 1952 Clubman's TT,*
*which was won by Eric Housley and in which*
*Bob McIntyre was 2nd, both at record speeds*
*with this re-designed engine.*

**1952 BSA C11G 247cc,** OHV,
single cylinder, re-built.
**£500–700**  *BCB*

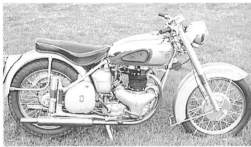

**1952 BSA A10 Gold Flash 650cc,** V-twin, original
specification, good mechanical and cosmetic order.
**£2,500–2,800**  *BKS*

**1952 BSA Bantam 125cc,** restored
to original condition.
**£750–900**  *BKS*

**1952 BSA D1,** good condition throughout, extensive
re-build, many new parts in the 1980s, finished in
the correct green and cream livery, furnished with
the optional plunger rear suspension.
**Est. £900–1,100**  *S*

**1952 BSA B31 350cc,** OHV,
single cylinder, restored.
**£1,700–2,000**  *BCB*

**1953 BSA B31 350cc,** plunger model.
**£1,500–1,900**  *AT*

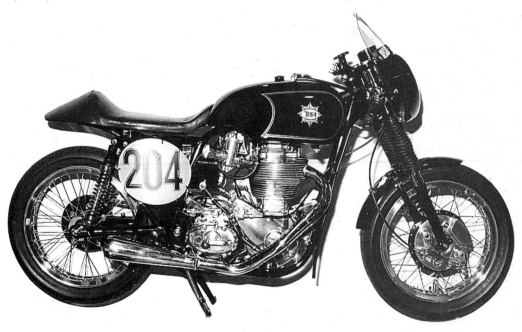

**1954 BSA CB34 Racer 499cc.**
**£4,500–5,500**  *GSO*

*This is the racing version of the*
*Gold Star Clubman.*

**1954 BSA D3 Bantam,** complete in all major respects, but in need of restoration.
**Est. £350–450** *S*

**1954 BSA D1 Bantam 125cc,** fully restored, good working order, green paintwork, original specification.
**£500–550** *S*

**1954 BSA BB32 Gold Star 350cc.**
**£4,500–5,000** *GSO*

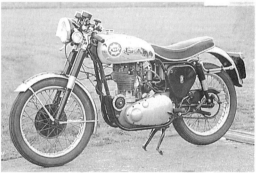

**1954 BSA Gold Star Clubman 500cc,** single cylinder engine to Clubman's specification, finished in black paintwork with chrome tank, black dual seat, good condition.
**£3,800–4,000** *S*

**1955 BSA Bantam 123cc,** 2-stroke, single cylinder.
**£200–300** *PS*

*This machine is basically the 148cc D3 model which has been fitted with a 123cc cylinder barrel and piston so that it can be used by a learner.*

**1955 BSA B34 Gold Star,** good paintwork, chrome and transmission, touring handlebars and footrests fitted, tachometer, swept back exhaust system, finished in black with a silver and chrome petrol tank, original log book.
**£7,500–8,000** *S*

**1955 BSA C101 250cc,** complete and original in all major respects.
**£480–520** *S*

**1955 BSA B31 350cc,** original specification, re-built magneto, new plug, HT lead, battery, and inner tube.
**£1,600–2,000** *S*

*l.* **1956 BSA M33 500cc,** B-series engine in M-series frame, big single.
**£2,000–2,500** *BLM*

*This motorcycle made a good sidecar workhorse.*

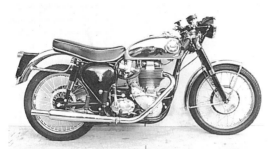

**1957 BSA Gold Star 350cc.**
**£4,500–5,000** *VER*

**1957 BSA B31 350cc,** Ariel brakes.
**£1,700–2,500** *BLM*

**1954 BSA D3 Bantam 125cc.**
**£600–800** *BLM*

**1957 BSA A7SS Shooting Star 500cc.**
**£3,000–4,000** *BLM*
*This is the sporting version of the standard A7.*

**1957 BSA D3 Bantam 150cc.**
**£400–600** *AT*

**1957 BSA B31 350cc,** restored to
original condition.
**£1,800–2,500** *SW*

**1958 BSA Gold Star 500cc,** good
condition, Clubman's trim, ex-race
history, now road used.
**£6,500–7,500** *SW*

**1958 BSA DB32 Gold Star 348cc,**
single cylinder, totally restored.
**£5,000–6,000** *BCB*

*l.* **1958 BSA Gold Flash,** extensively restored
to excellent condition, engine re-built, finished
to a high standard in black with the customary
chrome parts.
**£2,000–3,000** *S*

**1958 BSA DBD34 Gold Star 499cc,**
full Clubman's trim, Amal GP
carburettor, 190mm front brake.
**£9,000–10,000** *PC*

**1959 BSA Gold Flash 650cc.**
**£1,500–2,000** *PS*

*This Gold Flash has been restored. The tank top
panel grid is non-standard being usually seen on
Triumph motorcycles of the same era.*

**1959 BSA A10 Gold Flash 650cc,** finished in
bronze paintwork, with chrome tank and beige
dual seat, telescopic front forks, swinging arm
suspension, full lighting set and instruments,
good condition, features 4 forward speeds with
posi-stop gearchange.
**£2,500–2,800** *S*

**1959 BSA DB32 Gold Star 350cc.**
**£6,500–6,800** *BLM*

*The 350cc Gold Star motorcycles of this period
were regular winners in the Isle of Man
Clubman's race series.*

**1959 BSA Bantam Trials 125cc.**
**£600–700** *BMM*

**1960 BSA DB32 Clubman's Gold Star 350cc,**
good unrestored condition, old style log book.
**Est. £5,000–6,000** *BKS*

**1960 BSA Gold Flash 650cc.**
£3,000–4,000  *SW*

**c1966 BSA Victor Enduro 441cc,**
In good order throughout.
£1,500–2,250  *SW*

**1960 BSA DBD34 Gold Star 500cc,** RRT2
gearbox, steel tank, 1½in Amal GP carburettor,
in re-built excellent condition.
£7,000–8,000  *MR*

**1960 BSA A10 Super Rocket 647cc,** re-built.
£2,500–3,000  *BCB*

**1960 BSA CB34 500cc,** not original.
£4,000–5,000  *GSO*

**1960 BSA Gold Star Scrambler 499cc,**
original condition.
£3,500–4,000  *GSO*

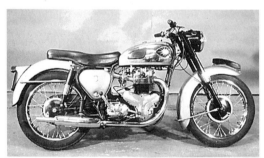

**1961 BSA A7 Shooting Star 497cc.**
£2,000–2,300  *S*

**1961 BSA A10 650cc,** to original
specification, with original log book.
£3,000–3,500  *BKS*

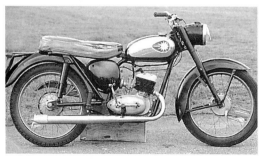

*l.* **1961 BSA D7 Bantam 175cc,**
finished in black, complete.
£200–250  *S*

**c1962 BSA B40 350cc,** ex-Home Office machine.
**£1,000–1,500** *BLM*

**1962 BSA C15 Star 250cc.**
**£600–700** *BLM*

*This model is the most popular 250 in the BSA Owners' Club.*

**1962 BSA A65 Star 654cc.**
**£2,800–3,000** *BLM*

*A smooth and reliable motorcycle with low compression pistons.*

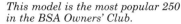
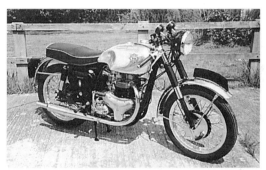

**1962 BSA A10 Rocket Gold Star 650cc Replica.**
**£3,000–3,500** *BLM*

**1962 BSA SS80 Sports Star 247cc,** good unrestored condition.
**£1,000–1,200** *BCB*

*The sporting version of the C15.*

**1962 BSA SS80 Sports Star 247cc.**
**£1,500–1,800** *PS*

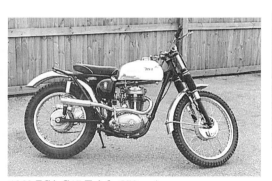

**1963 BSA C15 Trials 250cc,** in excellent condition.
**£1,500–2,000** *SW*

**1963 BSA SS90 350cc.**
**£1,000–1,500** *BLM*

*r.* **1964 BSA Victor Scrambler 441cc,** as new condition.
**£3,000–3,500** *SW*

*This motorcycle is used in Classic Moto Cross.*

**1964 BSA C15 250cc,** fully restored, current V5.
**£600–650** *BKS*

**1965 BSA B40,** original panniers, good general condition, V5 and original log book.
**Est. £800–1,000** *S*

**1965 BSA A65 Lightning 654cc,** OHV twin, non-standard hi-level pipes, restored to original condition.
**£2,500–3,000** *BCB*

**1965 BSA Bantam D7 Super 175cc,** finished in red and cream, very good restored condition.
**£450–550** *MR*

*The D7 Super 175 Bantam replaced the D5 in 1959 and ran through to the end of 1965, before being replaced by the D10 of the same capacity.*

**1965 BSA C15 250cc,** restored to as new condition.
**£650–800** *SW*

**1966 BSA A65 Lightning Rocket 650cc.**
**£2,700–3,000** *BLM*

**1966 BSA Barracuda 250cc,**
OHV single, restored.
**£700–950** *BCB*

**1967 BSA A65T Thunderbolt 650cc.**
**£2,000–2,300** *BLM*

*One of the most popular models in the BSA Owners' Club.*

*l.* **1967 BSA Bantam Bushman 175cc,** single cylinder, 2-stroke, bore and stroke 61.5 x 58mm.
**£1,700–2,200** *C(A)*

*Designed as an agricultural bike, the Bushman D10-B was only manufactured in 1966 and 1967 and together with the D14/4 and D175 represented the final form of the famous Bantam lineage. Over half a million Bantams were built between 1948–71.*

**1967 BSA Bantam 175cc,** finished with blue
mudguards and petrol tank, black frame and
forks and silver hubs, good overall condition.
**£400–450**  *S*

**1969 BSA B44 441cc.**
**£800–1,500**  *BLM*

**1969 BSA A65 Thunderbolt 654cc,**
OHV, big twin, tank, mudguards and
silencers not original, re-built.
**£1,200–1,500**  *BCB*

**1969 BSA D14/4 175cc,** 2-stroke single,
original condition.
**£400–600**  *BCB*

**1970 BSA B44,** chromium plated
fuel tank and Gold Star type
silencer, good running condition.
**£1,800–2,000**  *S*

**1971 BSA B50 Gold Star 499cc.**
**£1,400–1,800**  *BLM*

**1972 BSA Rocket 3 750cc.**
**£2,900–3,300**  *BLM*

**1972 BSA A65 Lightning 650cc.**
**£1,500–2,500**  *BLM*

*The oil-in-frame A65 had quite a
high seat position.*

*r.* **c1973 BSA A65 Thunderbolt
654cc,** OHV unit construction twin.
**£2,500–3,000**  *AtMC*

# BULTACO (Spanish)

**1965 Bultaco 125 TSS 125cc.**
**£4,000–5,000**  *S*

*The first water-cooled Bultaco, a factory development 125 TSS, made its debut at the Italian Grand Prix at Monza in September 1963, ridden by Johnny Grace. The production version appeared the following year. Water-cooling had the advantage of retaining power for the duration of the race, rather than slowing as the race progressed. This 1965 example comes complete with a Tommy Robb inspired fairing, which provided more rider protection than the factory original.*

**1962 Bultaco 196 TSS 196cc.**
**£3,000–4,000**  *S*

*Besides being fitted to its road and motocross models – plus a racing kart – the Bultaco company offered its 196cc engine as a road racer which first appeared in 1962. Only ever built in air-cooled form the 196 TSS featured a 4-speed close ratio gearbox, chain primary drive and maximum power was 31bhp at 10,000 rpm. It remained available until replaced by a 250 at the end of 1963.*

**1963 Bultaco 125 TSS 125cc.**
**£2,500–3,500**  *S*

*The first Bultaco racer was a works 125 in 1960. Works developments models appeared in 1961, followed by the production of 125 TSS in 1962. These early models all sported 4-speeds and air-cooling. In 1963 air-cooling was retained and 6-speeds introduced.*

# CALTHORPE (British)

**c1921 Calthorpe Villiers Lightweight 250cc.**
**£1,500–2,000**  *BLM*

# CHENEY-BSA (British)

**c1967 Cheney-BSA Scrambler 250cc,**
**£4,000–4,500**  *SW*

*This bike was a purpose-built Motocross machine.*

# CHENEY-TRIUMPH (British)

*l.* **c1963 Cheney-Triumph 500cc.**
**£2,200–2,600**  *BLM*

*This motorcycle was similar to machines campaigned by British team riders in International Six-Day Trials.*

# CLYNO (British)

**1916 Clyno 5/6hp Solo,** excellent V-twin side valve engine, chain drive, 3 forward speeds, complete with acetylene lighting and luggage carrier, new tyres, good condition.
**Est. £5,000–6,000**  *S*

*Taking their name from the inclined pulley patented by the company, Clyno motorcycles were made in Pelham Street, Wolverhampton from 1911 until 1923. During WWII Clyno supplied large numbers of motorcycles to the British and allied armies – including the Imperial Russian army.*

# COTTON (British)

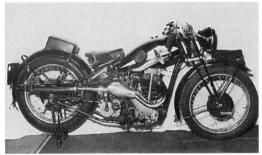

**1927 Cotton 500cc,** 4-valve.
**£7,000–8,000** *COEC*

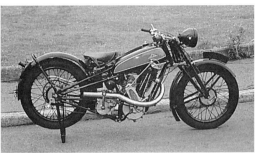

**1928 Cotton 500 495cc.**
**£3,000–3,750** *PM*

*The Gloucester based Cotton marque, started in 1920, was hallmarked by its frame. This dated back a further 7 years when Francis Willoughby Cotton first laid down his triangulated design that was to remain largely unchanged until 1939. This particular motorcycle is a 500 model with a Blackburne side valve engine.*

**1932 Cotton Lightweight 349cc,** JAP engine.
**£4,500–5,000** *COEC*
*Known as Lightweight for tax purposes.*

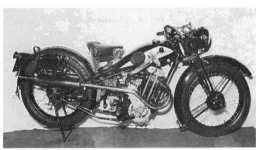

**1937 Cotton Sports 349cc,** JAP engine.
**£7,000–8,000** *COEC*

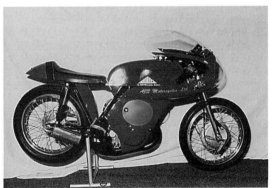

**1964 Cotton Telstar 250cc,** single cylinder fitted with Villiers Starmaker engine, 6-speed gearbox, MP forks and Grimeca front brake.
**£4,500–5,000** *COEC*

**1961 Cotton Continental 250cc,** Villiers 2T 2-stroke twin engine, totally restored to original specification.
**£2,000–2,500** *COEC*

**1974 Cotton Trials 170cc,** Minerelli engined 2-stroke, good condition, roadworthy, paintwork reasonable.
**£600–700** *S*

**1968 Cotton Trials 250cc,** Villiers engine.
**£1,000–1,400** *BLM*

# COVENTRY (British)

**1936 Coventry Eagle 250cc,**
OHV JAP engine.
**£1,000–1,500** *BLM*

**1932 Coventry Eagle 200cc,** Villiers engined
2-stroke, fully restored to original condition.
Est. £2,400–2,600 *S*

# DOT (British)

**1921 DOT 2¾hp,** JAP engine.
**£2,000–2,500** *DOT*

*The earliest known example of this machine
and the only belt-driven survivor.*

**1955 DOT TDM X4D,** Villiers 197cc
2-stroke engine, 4-speed gearbox, hydraulic
rear suspension, mechanically good, engine
re-built, transmission, gearbox, paintwork,
electrics, upholstery and tyres excellent,
the silencer is not original.
**£900–1,000** *S*

# DMW (British)

**1964 DMW Racing Hornet 250cc,**
mechanically 'race ready', non-standard
Yamaha 2LS front brake and MP forks, with
good paintwork in green and silver, good tank
and frame.
**Est. £2,800–3,500** *S*

*DMW motorcycles took their name from the
Dawson Motor Works, founded in 1945 to
initially build grass track racing machines.*

*r.* **1955 DOT SCH
Scrambler 197cc,**
Villiers 9E engine,
restored.
**£1,000–1,200** *PC*

**1960 DOT Works Trials 246cc,** Villiers single
cylinder 2-stroke engine, 4-speed gearbox, with
leading link forks frame, conventionally sprung
rear fork, new condition throughout, finished in
red with alloy petrol tank and mudguards.
**£1,500–2,000** *S*

**1961 DOT Demon Scrambler 246cc.**
**£1,500–2,000** *DOT*

*This machine was purchased in 1989 in scrap
condition. It has been driven on the road with
a sidecar.*

# DOUGLAS (British)

**1907 Douglas.**
**Est. £5,500–6,500** *LDM*

*This machine was found in a packing case in 1970. It was sold for £1,000 in 1975 and won best original bike in the Bristol Show 5 times.*

**1910 Douglas,** mint condition.
**Est. £5,000–5,500** *LDM*

*This machine is the same type as those taking part in the 'End to End' run in 1910.*

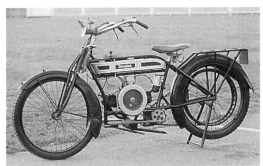

**1914 Douglas Solo 2¾hp 350cc,** original machine, has been the subject of an older restoration, complete with Gamage Nulite acetylene headlamp, luggage carrier and acetylene generator.
**Est. £4,000–5,000** *S*

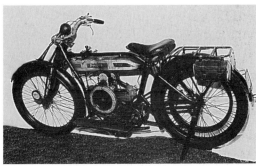

**1915 Douglas V 2¾hp,** largely unrestored, used regularly.
**£2,750–3,500** *LDM*

---

| Miller's is a price GUIDE |
| not a price LIST |

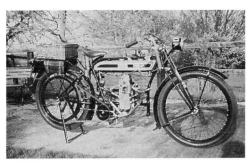

**c1921 Douglas 2¾hp 350cc.**
**£4,000–4,500** *BLM*

**c1922/3 Douglas 2¾hp,** horizontally opposed twin cylinder engine, 2-speed gearbox, complete with acetylene lighting set, luggage grid and leather toolboxes, flat tank, stand and footboards, usable.
**Est. £2,800–3,000** *BKS*

**c1924 Douglas EW 350cc.**
**£3,000–3,400** *BLM*

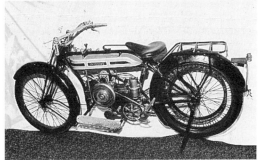

**1924 Douglas 2¾hp,** restored engine, in original condition.
**£3,000–3,500** *LDM*

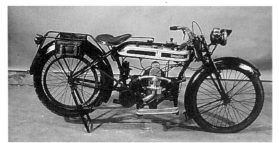

**1924 Douglas 2¾hp Solo,** painted in traditional Douglas livery of black with blue and white tank, acetylene lighting set, luggage grid, footrests and stand, good condition.
**£2,800–3,000** *S*

*The transmission of power is via primary chain to the gearbox with hand control gearchange, then by belt to the rear wheel.*

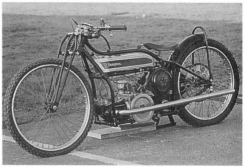

**c1928 Douglas Speedway DT,** semi-automatic gear mechanism, in ready to race condition.
**Est. £8,500–9,000** *S*

*This may have been an ex-works machine.*

**1948 Douglas OHV Twin 346cc.**
**£2,200–2,500** *BLM*

**1949 Douglas Mark III 348cc,** pan saddle, separate pillion seat, full lighting set and sprung front mudguard, blue and black paintwork.
**Est. £1,600–1,900** *S*

**1951 Douglas Plus 90 Special 350cc,**
4-stroke engine with 2-stroke induction via rotary valves, limited edition sports model.
**£15,000–16,000** *LDM*

**1952 Douglas MKV 350cc,** restored, mostly original, polychromatic blue.
**£2,800–3,000** *LDM*

**1955 Douglas Dragonfly 348cc.**
**£1,500–2,000** *PM*

*In 1955 Douglas introduced a new flat-twin machine. At first this was called the Dart, but by the time it reached production it had become the Dragonfly. The engine was basically as before, but most components had been revised, including stronger crankcase and crankshaft. The main changes were to the cycle parts which included a new duplex frame, an unusual fuel tank and headlamp mounting. Production ceased in March 1957.*

# DREADNOUGHT *(British)*

**1915 Dreadnought 250cc,** Villiers engine.
**£2,500–3,000** *BLM*

*A rare lightweight flat tanker, this model is believed to be the only one on the VMCC register of machines.*

# DUCATI *(Italian 1946–)*

Ducati's dominance of the world superbike racing in the 1990s is miles apart from its 1930s beginning as a radio manufacturer. During WWII the company supplied military equipment and was only able to restart radio production through a financial package with the Italian government and the Vatican each taking a 50% share in the Bologna firm.

The basis of this rebirth was a 49cc 4-stroke engine unit, which was created by Aldo Farinelli. Huge numbers of this micro-engine were produced during a ten year period beginning in 1946, and the success of this venture was to pave the way for Ducati's entry into the manufacture of complete motorcycles, the first of which made its debut in 1950. This machine was powered by a 60cc version of the Cucciolo (Little Pup) micro-engine and this was followed by 65, 98 and 124cc OHV singles.

In 1954 Ducati's management signed up the engineer Fabio Taglioni and the start of a legend was born. Taglioni – or 'Dr. T' as he is known to countless enthusiasts – was not only a talented designer, but also a racing enthusiast. The following year saw Taglioni's first design for his new employers, this

being the 100 Gran Sport; a machine which was to be the beginning of a vast line of OHC singles culminating with the 450 model in 1969.

Next came a 125 single in both valve spring and Desmo form. The latter winning the first race it contested, the 1956 Swedish Grand Prix.

By 1958 Ducati not only had a new range of sports roadsters headed by the 175 Sport and 200 Elite, but also finished the year in second place in the world championships, after a season-long battle with MV Agusta. In 1959 a youthful Mike Hailwood won his first ever Grand Prix on a Ducati (the 125cc Ulster) and special 250 and 350 twins were commissioned by his father Stan for Mike to race.

The 1960s saw the introduction of the famous 250 Mach 1. However, the company was in trouble and the government bought out the Vatican's remaining shares by the end of the decade. A significant business partnership began in 1983 when Ducati began supplying engines to Cagiva. Two years later Cagiva – run by the Castiglioni family – bought the Ducati business.

**1965 Ducati Mach 1 248cc.**
**£4,000–4,500**  *PC*

*Seen by many as Ducati's best single, the 'racer-on-the-road', the Mach 1 was introduced to an amazed public in September 1964. With a maximum of just over 100mph it was the fastest roadster 250 at the time of its launch. With a large SS1 29 Dell'Orto carburettor, 10.5:1 Borgo forged piston and 5-speed gearbox its specification read more like an exotic racer than a mere street bike. This pristine example has been carefully restored to its original glory by an American enthusiast.*

*l.* **1966 Ducati Monza Junior 156cc,**
OHC single, 4-speeds, 16in wheels, fully restored by Scottish owner.
**£1,000–1,400**  *PC*

**1963 Ducati Daytona 250cc,** non-standard single seat, British headlight and megaphone silencer fitted, toolboxes are from the later Mach 1 model, excellent compression and runs well.
**£2,500–3,000** *BKS*

**1971 Ducati Monza Junior 156cc,** mechanically and cosmetically in good condition.
**Est. £800–1,000** *S*

- **The Pantah was the first machine to boast the trellis frame and belt drive motor, two features which remain on current models.**
- **The 600cc version of the Pantah took Tony Ruffer to no less than four world F2 titles in the early 1980s.**

**1974 Ducati 450 Street Scrambler.**
**£1,400–2,400** *S*

*The final versions of Ducati's popular 'hunting and fishing' 450 Street Scrambler models were built in 1974 and utilised the double-sided drum as found on the roadster Mark 3 models. Forks were heavyweight 35mm Marzocchi's, whilst both centre and side stands were part of the specification.*

**1971 Ducati Silver Shotgun 248cc,** single cylinder Desmo, 4-stroke, full race tuned, racing tank, seat and mudguards, Dell'Orto pumper carburettor, Ceriani GP racing front forks and a massive double-sided 2 leading shoe front brake.
**£5,600–6,600** *C(A)*

*This particular machine was modified in Australia where there has been a long-standing love affair with Ducati single cylinder models. It was produced in 1971 as a 250 Desmo and has been converted into a road-going replica of the works model raced by the factory in the late 1960s by Bruno Spaggiari.*

**1972 Ducati Mick Walker Special 340cc.**
**£4,000–5,000** *PC*

*Built in both narrow and widecase versions, the latter with the addition of 450cc, the Mick Walker framed Ducati singles were an attractive and affordable adaptation of Ducati's classic bevel-drive single during the 1970s. They were equally at home on both short circuits and the Isle of Man TT circuit. This example is the original prototype as raced by Dave Arnold and Dave Street and later campaigned during the 1980s by classic racer Jim Porter. It features a narrowcase 340cc 5-speed engine.*

**1974 Ducati 750SS Desmo 748cc,** original except 2 into 1 exhaust.
**£11,000–12,000** *GLC*

*One of only 200 road-going replicas of the 1972 Imola 200 race winner.*

*l.* **1979 Ducati Darmah SD 900,**
OHV 90 degree V-twin.
**£2,000–2,500** *PC*

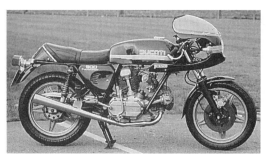

**1980 Ducati 900SS,** re-built engine and transmission, professionally resprayed, twin headlamp, half fairing, good condition.
**£3,500–4,000** *S*

# DUNSTALL *(British)*

**1978 Dunstall F1 Suzuki Racing**
1,000cc, 4-cylinder, 4-stroke.
Est. **£3,600–3,900** *S*

# ENFIELD *(Indian)*

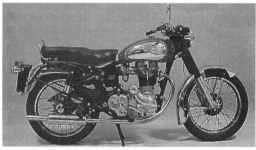

**1978 Enfield-India 350 Bullet 346cc,** single cylinder OHV, 4-stroke, bore and stroke 70 x 90mm, 18bhp.
**£1,800–2,200** *C(A)*

*Enfield-India was originally a branch of Royal-Enfield of Redditch. When the parent company ceased production Enfield-India continued with what was basically an old but still very popular design. Extensively used by military and civilian authorities in India, many are also still exported to enthusiasts around the world.*

**1920 Excelsior 2¾hp,** JAP side valve, single cylinder engine, Burman 2-speed gearbox, primary chain to belt final drive, brown leather saddle, black paintwork, excellent condition
**£4,200–5,000** *S*

**1980 Ducati 900 Mike Hailwood Replica 864cc,** heavy duty clutch conversion, 2 into 1 exhaust, very good condition.
Est. **£5,000–6,000** *BKS*

*The Mike Hailwood Replica was launched in 1979 as a limited series of 500 (of which 200 were imported into the UK) to commemorate Mike Hailwood's remarkable TT comeback the previous year on a Sports Motorcycles prepared Ducati.*

# EMW *(German)*

**c1952 EMW 350cc,** OHV single.
**£2,000–2,500** *AtMC*

*East German version of pre-war BMW.*

# EXCELSIOR *(British)*

**1936 Excelsior Manxman 250cc.**
**£3,500–4,000** *PM*

*The first of Excelsior's famous Manxman models were produced in 1935, in 246cc and 349cc sizes. These were followed in 1936 by the 496cc F14 model. All were listed in sports and racing forms and did much to lift Excelsior into the realms of the prestige motorcycle builder.*

**1957 Excelsior 198cc**
**£100–200** *BMM*

# FB MONDIAL *(Italian)*

**1953 FB Mondial Racing 125cc,** good condition.
**Est. £17,000–18,000**  *BKS*
*This machine is the customer version of the factory's world championship winning racer.*

# FRANCIS-BARNETT
*(British)*

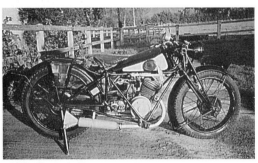

**1930s Francis-Barnett 250cc.**
£1,500–2,000  *BLM*

**1954 Francis-Barnett 62 Falcon 197cc.**
Villiers 7E engine and 3-speed gearchange.
**£800–1,200**  *BLM*

**1960 Francis-Barnett 87 Falcon 247cc,**
AMC engine.
**£300–400**  *PM*

# FN *(Belgian)*

**1948 FN X111 450cc.**
**£1,600–1,800**  *BLM*
*This is a rare and collectable machine.*

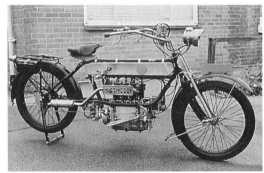

**1914 FN 748cc,** 4-cylinder classic.
**£14,000–15,000**  *AtMC*

**1958 Francis-Barnett 81 Falcon 197cc,**
Villiers 10E engine, period set of legshields, in need of restoration.
**Est. £300–400**  *S*

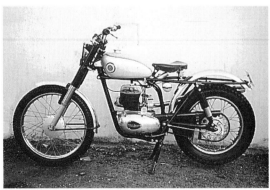

**1959 Francis-Barnett 85 Trials 249cc,** road registered, in excellent restored condition.
**£700–1,100**  *MR*

**1968 Aermacchi Ala d'Oro 248cc.**
**£7,000–8,000** *S*

*By 1968 Aermacchi was offering its horizontal pushrod singles in both 248cc and 344cc engine sizes, both short-stroke.*

**1923 AJS 348cc,** Anderson steering damper, leather fronted tool boxes, full acetylene lighting set, contemporary horn, nearly original flat tanker and in good condition.
**£3,000–4,000** *S*

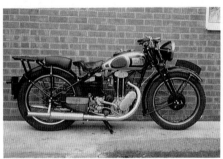

**1936 AJS Model 22 250cc,** restored to a roadworthy condition.
**£2,000–3,000** *NAC*

*Very few examples of this year / model have survived. This is a 1935 model, registered in 1936 and is one of the company's first models with footchange gearbox.*

**1951 AJS 7R 'Boy' Racer 349cc,** 75.5 x 78mm bore/stroke, single cylinder, OHC, 4-stroke, 30bhp.
**£23,500–25,000** *C(A)*

*Formerly the pre-war R7 model, it reappeared as the 7R in 1949. Built for racing only, by the end of the 1958 season, 524 of these stylish and reliable machines had been sold.*

**1951 AJS 18S 500cc,** overhead valve single, fitted with AMC engine, telescopic forks, swinging arm rear suspension, in good condition.
**£2,000–3,000** *S*

**1951 AJS 16MS 349cc,** OHV single cylinder model, rebuilt and in good condition.
**£1,100–1,800** *PS*

**1954 AJS Jam Pot Model 350cc.**
**£900–1,100** *PS*

**1960 AJS AMC G12 CSR 650cc,** twin competitive engine, modified into a Clubman's racer with seat, handlebar and other modifications.
**Est. £1,600–2,000** *S*

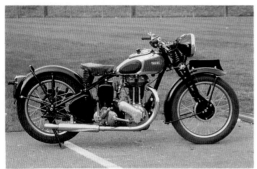

**c1938 Ariel Red Hunter 350cc,** girder forks with a rigid frame, single cylinder overhead valve engine, Burman 4-speed gearbox, in good condition.
**£1,500–2,000** *S*

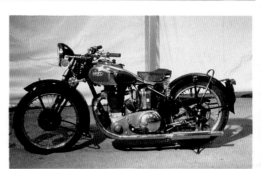

**1939 Ariel Red Hunter 500cc,** single cylinder overhead valve 4-stroke engine, separate 4-speed gearbox, girder forks with coil spring suspension and rigid frame at rear.
**£3,000–4,000** *S*

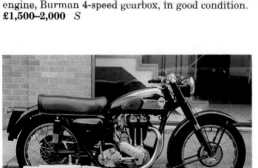

**1955 Ariel NH Red Hunter 346cc,** OHV single and restored.
**£1,750–2,000** *BCB*

*Built from 1945–1959 this 1955 model has separate pushrod tubes unlike later models.*

**1929 Ariel Brooklands Sprint 500cc,** small racing petrol tank, dual chamber carburettor, Albion gearbox, racing handlebars and fishtail exhaust.
**£4,800–5,200** *BKS*

**1955 Ariel VH 499cc,** single cylinder with Burman GB6 gearbox, fully restored to original trim.
**£2,300–2,500** *ABT*

**1953 Ariel NH Red Hunter 347cc,** 72 x 85mm bore/stroke, single cylinder, OHV single port, 4-stroke, 14.2bhp.
**£1,900–2,100** *C(A)*

**1956 Ariel Huntmaster 646cc,** OHV twin cylinder, rebuilt with full width brake hubs and headlamp cowl.
**£1,750–2,000** *BCB*

**1957 Ariel Red Hunter 346cc.** Est. **£800–900** *S*

*This model has been in dry store for the past twenty-five years, but is complete and running.*

**1958 Ariel HT5 Trials 497cc.**
**£4,200–4,800** *S*

*The HT5 was Ariel's Trials version of the Red Hunter. It was available from the 1956 season until the decision was taken at the end of the 1950s to concentrate on twin 2-strokes.*

**1958 Ariel Colt 198cc,** OHV engine, separate gearbox with typical BSA pressed steel chaincase.
**£600–800** *AT*

**1959 Ariel 247cc,** inclined twin 2-stroke engine, pressed steel frame with full enclosure of mechanical elements, motorcycle wire wheels.
**Est. £350–450** *S*

**1961 Ariel Golden Arrow 247cc,** 2-stroke twin engine, fully restored and in good condition.
**Est. £2,250–2,750** *S*

**1960 Ariel Leader 247cc.**
**£420–520** *PS*

**1962 Ariel Golden Arrow 247cc,** twin cylinder 2-stroke engine.
**Est. £1,000–1,200** *BKS*

**1964 Ariel Arrow 247cc,** 2-stroke engine, frame with pressed steel box member, extending from steering head to rear suspension, restored.
**Est. 1,250–1,500** *S*

**1964 Ariel Arrow 247cc,** 2-stroke twin cylinder.
**£300–400** *PS*

**1961 Bianchi Tonale 175cc,** 4-stroke, OHV engine.
**£800–900** *S*

*Tonale was at the top of the range with its chain driven 175cc overhead motor. This example dates from 1961, and has matching frame and engine numbers.*

**1973 BMW R50/5 498cc,** good condition.
**£1,500–2,500** *BKS*

**1927 Brough Superior 680cc,** JAP overhead valve engine, Miller lighting, Bonniksen time and speed meter.
**£10,000–12,000** *S*

**1933 Brough Superior SS80 980cc,** V-twin side valve JAP engine, 4-speed Sturmey-Archer gearbox, Castle forks, in good running order.
**Est. £7,500–8,500** *S*

**1959 BMW R69 590cc,** OHV horizontally-opposed twin engine, in good condition.
**£1,100–2,100** *PS*

**1969 BMW R60 597cc.**
**£2,800–3,200** *S*

**1930 Brough Superior Solo 680cc,** finished in black with gold lining, in good condition.
**£12,500–13,500** *S*

**1937 Brough Superior 1150STD,** rebuilt with deviations from standard specification, including lever-operated advance/retard mechanism.
**£9,500–10,500** *S*

**r. 1939 Brough Superior SS80 990cc,** 84 x 90mm bore/stroke, Matchless V-twin, side valve 4-stroke, 32bhp.
**£12,000–13,000** *C(A)*

**1930 BSA Sloper 500cc.**
**£3,000–3,800** *AT*

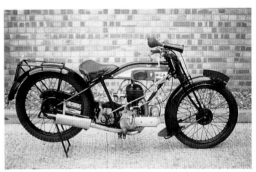

**1928 BSA H28 557cc,** single
cylinder, vertical engine unit.
**Est. £2,000–2,500** *BKS*

**1934 BSA B1 250cc,** side valve single, restored to
original condition.
**£1,250–1,500** *BCB*

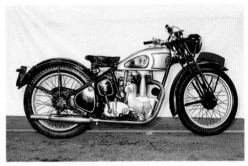

**1938 BSA B21 249cc,** OHV single
cylinder, original and in good condition.
**£1,600–2,000** *PS*

**1941 BSA M20 499cc,** side valve single,
rebuilt to original specification.
**£800–900** *BCB*

*M20s were one of the British Army's main
Despatch Rider bikes of WWII.*

**1951 BSA Bantam 123cc,** finished in
correct green livery, complete with
lighting set and luggage carrier,
original.
**£300–400** *S*

**1952 BSA C10L Deluxe 246cc,** side valve single,
restored to original condition.
**£700–850** *BCB*

**1955 BSA M21 600cc,** side valve.
**£950–1,150** *AT*

*Later engined version of the M20, the
600cc M21 was able to cope with the
fitment of a sidecar.*

**1957 BSA Shooting Star 498cc,** twin cylinder,
totally restored.
**£2,500–3,000** *ABT*

*This sporting version of the BSA pre-unit A7
498cc twin, was largely overshadowed by the
similar, but more powerful 646cc Road Rocket.*

**1958 BSA C12 246cc,** OHV single,
rebuilt to original condition.
**£750–1,000** *BCB*

**1961 BSA B40 343cc,** OHV, single cylinder, totally
restored.
**£1,200–1,500** *BCB*

**1961 BSA A10 647cc,** twin cylinder,
in excellent condition.
**£2,500–3,000** *ABT*

*A good example of the A10 and is a
model which helped BSA maintain their
good reputation.*

**1967 BSA A50 Royal Star 499cc,**
totally restored.
**£2,300–3,000** *PS*

*The A50 Royal Star was listed
from 1963–1969. Its 499cc engine
produced 32bhp at 6250rpm.*

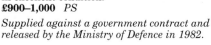

**1962 BSA Rocket Gold Star 647cc,**
OHV twin, restored.
**£6,000–7,000** *BCB*

**1967 BSA B40 343cc,** OHV single cylinder,
in excellent condition.
**£900–1,000** *PS*

*Supplied against a government contract and
released by the Ministry of Defence in 1982.*

**1967 BSA A65 Lighting 654cc,**
OHV twin, restored.
**£2,500–3,000** *BCB*

**1969 BSA Thunderbolt 654cc,**
OHV twin, restored.
**£2,500–2,750** *BCB*

**1970 BSA Firebird Scrambler 654cc.**
**Est. £2,500–2,800** *S*

*This machine, produced almost entirely for*
*the American market, was described as a*
*street scrambler.*

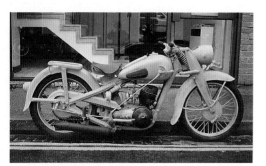

**1940 DKW Military 349cc,** 2-stroke, in original
unrestored condition.
**£500–700** *BCB*

**1968 Bultaco 125 TSS 125cc.**
**£3,000–4,000** *S*

*By 1968 Bultaco's full range of bikes*
*consisted of 21 models, with capacities of*
*75, 100, 125, 175, 200, 250 and even a 360*
*motocrosser, a far cry from the solitary*
*Tralla 101 roadster of a decade before.*

**1957 DMW 225cc,** Villiers 2-stroke single
cylinder, fully restored to original specification.
**Est. £1,500–1,900** *S*

**1961 DMW Dolomite 249cc,** Villiers
2-stroke twin cylinder, 4-speed gearbox, full
width brake hubs and Earles type front forks.
**£750–1,000** *BCB*

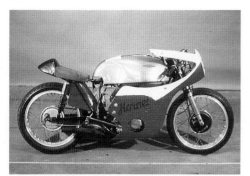

**1965 DMW Hornet Solo Racing 247cc.**
**£2,500–3,000** *S*

*In October 1962 the Villiers company*
*announced their new 247cc Starmaker engine.*
*DMW was quick to follow with the launch of a*
*Starmaker powered road racer in 1963. This*
*was named the Hornet.*

**1964 DMW Hornet Racing 247cc,** 6-speed
gearbox, race prepared to original specification.
**Est. £2,500–3,000** *BKS*

**1958 DOT RCA 349cc,** P. Hogan designed twin cylinder, Villiers 6-speed gearbox, race prepared condition.
**Est. £1,000–1,500** *S*

*l.* **1951 Douglas Mk5 349cc,** carefully restored.
**£2,250–2,750** *PM*

**1975 Ducati 750 Sport.**
**£5,000–5,995** *GLC*

*The first Ducati V-twins were produced in the early 1970s. In 1972 came the famous Imola victory and the launch of the 750 Sport. This model was one of the last made.*

**1954 Excelsior Consort 98cc,** 2-speed Villiers engine.
**£300–350** *NAC*

*Excelsior were one of the earliest British motorcycle manufacturers, producing their first model in 1896. This example was a 'barn discovery'.*

*l.* **1959 Excelsior 98cc,** 2-stroke Villiers single power unit.
**£200–300** *S*

**1966 Ducati Cadet 100,** original condition.
**£800–1,000** *PC*

*Little remembered today, the legendary Ducati marque offered a large range of 2-stroke models in the 1960s.*

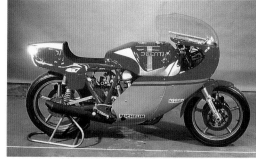

**1961 Ducati Daytona 248cc.**
**£1,750–2,250** *PM*

*With the Daytona (known as the Diana outside Britain) Ducati were able to offer a machine which could challenge anything in its class.*

**1978 Ducati NCR 864cc.**
**£20,000–22,000** *S*

*This is one of a small batch of specially constructed 900FI models built for Ducati by NCR.*

**c1973 Ducati 750S/SS,** rebuilt.
**£4,600–5,000** *S*

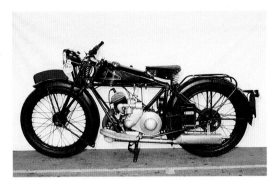

**1934 Francis-Barnett Lapwing,** 2-stroke single cylinder, unrestored and in good working order. **£800–1,000** *PS*

**1955 Francis-Barnett Plover 150cc,** Villiers single 2-stroke engine, rebuilt. **£300–400** *BCB*

**1966 Greeves RDS Silverstone 246cc.** **£2,500–3,500** *S*

*After the RBS came the RCS in 1965 and then on to the model shown here, the RDS in 1966, considered by most observers to have been the best of all the 250 Greeves tarmac racers.*

**1961 Greeves Sports Twin 249cc,** Villiers 2T engine, alloy front frame section, Siamesed exhaust, in original unrestored condition. **£700–1,000** *BCB*

**1942 Harley-Davidson WLA Military,** 45 cu in engine, in original condition. **£12,500–13,500** *Gen*

**1942 Harley-Davidson WLA Civilian 740cc,** V-twin, side valve, 4-stroke, 23bhp. **£8,000–9,000** *C(A)*

**1947 Harley-Davidson FL 74 Knucklehead,** rebuilt, in excellent condition. **£12,500–13,500** *Gen*

**1979 Harley-Davidson Sportster 1000cc.** **£5,000–5,500** *S*

*This machine was imported from the USA in October 1993.*

**1964 Honda CB160 Racer**, racer kitted with engine capacity increased to 180cc.
**£2,000–2,500** *VMCC*

**1970 Honda CB750 748cc**, 4-cylinder, air-cooled engine.
**£4,500–5,500** *S*

**1972 Honda CB175**, twin carburettors, 5-speed gearbox, electric starter.
**£750–800** *S*

*The CB175 was the sports' version of the CD175 OHC twin. This is the original version.*

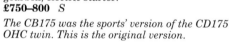

**1970 Honda CB450 450cc**, OHC vertical twin engine, in original and excellent condition.
**£2,000–2,500** *PS*

**1973 Honda SL125 122cc**, overhead camshaft, single cylinder engine, 5-speed constant mesh gearbox, in excellent condition.
**£1,500–1,800** *S*

**1975 Honda CB175 K6 175cc.**
**£350–450** *S*

**1967 Honda Z50M Monkey Bike.**
Est. **£1,200–1,500** *S*

*One of Honda's more unusual models, it soon became a regular sight at race meetings as paddock bikes and at resorts as tenders to yachts.*

**1972 Honda ST70 72cc.**
**£300–400** *S*

*This machine was a larger version of the original Monkey Bike, later called the Dax.*

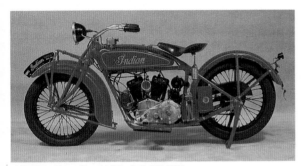

**1925 Indian Scout Model GE 596cc,**
V-twin, side valve, 4-stroke duplex frame and
leaf spring forks.
**£5,500–6,500**  *C(A)*

**1946 Indian Chief Solo,** in good
and original condition.
**Est. £10,000–12,000**  *ALC*

**1959 Itom 50cc.**
**£1,500–2,500**  *S*

*The first Itom was a cyclemotor which was
designed in 1944. Based in Turin the Company
soon built up an excellent reputation for its
50cc and and later 65cc models, all single
cylinder piston-ported 2-stroke engines.*

**1953 James Comet 98cc,** all stainless
steel including cantilever.
**£1,000–1,500**  *BIR*

**1950 James Comet 100 99cc,**
2-speed Villiers engine, girder forks
and rigid frame.
**£400–500**  *NAC*

*The James Comet is best referred to as
'very basic transport' with a 99cc engine.*

**1957 James L15 Cadet 147cc,** 2-stroke
single cylinder.
**£300–350**  *PS*

**1974 Kawasaki Model H1 MkIII,**
498.8cc 3 cylinder 2-stroke engine,
excellent condition.
**Est. £1,200–1,500**  *S*

**1975 Laverda Jota 980cc,** 3-cylinder
overhead camshaft, unrestored.
**£2,000–2,500**  *BCB*

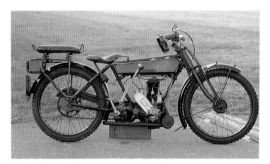

**1926 Levis 247cc.**
**£1,200–1,500** *S*

*Levis machines were made in Stechford, Birmingham, from 1911–39 and were once a leading name in the British motorcycle industry.*

**1934 Levis Model B34 247cc.**
Est. **£1,500–2,000** *HOLL*

*By 1933 the Levis 4-strokes had the option of 4 speeds, but was otherwise largely unchanged from earlier years.*

**1937 Levis Model L 347cc.**
Est. **£1,200–1,500** *HOLL*

*In addition to the Special OHV machines, a trio of Light models were added in 1937.*

**1939 Levis Model SV 346cc,** alloy head, exhaust lift cable.
**£1,500–2,000** *HOLL*

*New for 1939 was a 346cc single with side valve engine, which was Levis' first and only model with this configuration.*

**1957 Maico Taifun 394cc,** split crank case top and bottom, helical gear primary drive and fully enclosed duplex chain rear drive.
**£1,800–2,500** *MOC*

*There were only 40 of these machines ever imported.*

**1939 Levis Trials 346cc.**
**£1,000–1,500** *HOLL*

*There were competition versions of both the 346 and 498cc Levis singles available at the end of the 1930s.*

**1957 Maico Taifun 394cc,**
fully restored to original condition.
**£1,800–2,500** *S*

*The Taifun (Typhoon) was produced in 348 and 394cc engine sizes.*

**1929 Magnat-Debon Terrot BM 350cc,**
in good condition.
**£1,500–2,000** *S*

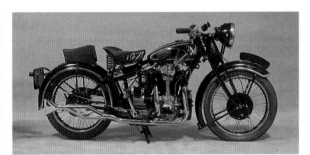

**1931 Matchless Silver Hawk Model B 593cc,**
V4, OHC, 4-stroke engine.
**£8,000–10,000** *C(A)*

**1955 Matchless Model G9 497cc.**
**£1,000–1,500** *PM*

**1951 Matchless G9 Twin 497cc.**
**£2,500–3,000** *AT*

**1959 Matchless Model G80 497cc.**
**£1,200–2,000** *PS*

*This is probably the best of all the AMC single cylinder roadsters of the era. The Matchless G80 and its AJS brother, the Model 18, were well respected for their solid dependability.*

**1960 Matchless G80 Special,** Norton twin
leading shoe front brake, single seat, 'Goldie'
pattern silencer and chrome mudguards.
**Est. £4,300–5,000** *S*

**1961 Matchless Model G12 646cc,** OHV twin
engine, restored to original concours condition.
**£2,500–3,000** *BCB*

**1958 Matchless Model G11 646cc,**
OHV twin cylinder, rebuilt.
**£1,750–2,250** *BCB*

**1953 Matchless G3LS 350cc,** overhead valve
single cylinder 4-stroke engine, 4-speed gearbox,
telescopic forks, to original specification.
**£1,200–1,800** *S*

*r.* **1958 Matchless G80S 497cc,**
single cylinder overhead valve engine,
in original condition.
**£1,500–2,500** *S*

**c1939 Moto Guzzi Albatros Racing 250cc,** Moto Guzzi engine with exposed valve springs and outside flywheel, girder forks.
**Est. £7,200–8,000** *S*

**1950 Moto Guzzi Airone Sport 498cc,** horizontal single cylinder engine layout, telescopic forks and sprung frame.
**Est. £2,800–3,400** *S*

**1956 MV Agusta Mondella Sport 172cc,** overhead camshaft and 4-speed gearbox.
**£2,500–3,500** *S*

**1955 Moto Guzzi Airone Turismo 247cc,** single horizontal cylinder and outside flywheel.
**£2,500–3,000** *S*
*This motorcycle was imported from Italy.*

**1980 Morini 3½ Strada 344cc.**
**£1,000–1,500** *S*
*The Morini 3½ Strada, first introduced in 1973, had by 1980 evolved with twin front discs, a black exhaust system and square CEV headlamps.*

**1931 New Hudson Standard Model 3 496cc,** bore/stroke 79.5 x 100mm, single cylinder, OHV, 4-stroke engine.
**£3,500–4,500** *C(A)*

**1937 New Imperial 500cc,** New Imperial sloping pushrod, OHV twin-port engine, girder forks with rigid frame at rear, good condition throughout.
**£1,500–2,500** *S*

**1957 Nimbus 746cc,** bore/stroke 60 x 66mm, 4-cylinder in-line OHC, 4-stroke engine.
**£4,000–5,000** *C(A)*
*Fisker & Nielsen Ltd. Copenhagen, Denmark's largest motorcycle factory, produced this model from 1920–57.*

**1927 Norton Model 18 Sports 490cc,** cylinder dimensions 79 x 100mm, OHV, new brake shoes, tubes and original beaded edge rim wheels.
**£6,000–8,000** *S*

**1931 Norton Model 18 500cc,** rebuilt.
**£3,000–3,500** *BCB*

**1936 Norton 348cc,** finished in correct black and silver Norton livery.
**£1,500–2,000** *S*

**1951 Norton 500T Trials 490cc,** all alloy engine, shortened version of the diamond 16H frame, finished in silver with a black frame.
**Est. £3,300–3,800** *S*

**1954 Norton International ES2 490cc,** single overhead valve engine, frame replaced by featherbed frame, mechanically in good condition.
**£4,000–5,000** *S*

**c1952 Norton 30M Manx 499cc,** gearbox overhauled, in original livery, in good condition.
**Est. £13,000–14,000** *S*

*This machine is an ex-Francis Beart racing motorcycle.*

**1957 Norton Model ES2 490cc,** OHV single, in original unrestored condition.
**£1,400–1,800** *BCB*

**c1959 Norton Manx 30M Double Knocker Racing 499cc,** double overhead camshaft engine, AMC 5-speed gearbox, twin spark ignition, rebuilt.
**£19,000–20,000** *S*

**1967 Norton Model 650SS 646cc,** restored and in good condition.
**£2,800–3,200** *BCB*

**1970 Norton Commando Fastback 745cc,** OHV twin, not original mudguards, rebuilt.
**£2,000–2,500** *BCB*

**1973 Norton Commando 850cc,** OHV twin cylinder, restored.
**£2,750–3,250** *BCB*

**1971 Norton Norvil Commando 'Yellow Peril' Production Racer 745cc,** OHV, twin cylinder engine.
**£2,000–2,500** *BCB*

**1937 Panther Model 100 600cc,** OHV single twin port, rebuilt.
**£1,750–2,000** *BCB*

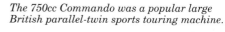

**1973 Norton Commando Interstate 745cc,** 5-speed gearbox. in good overall condition.
**£2,000–2,500** *S*

*The 750cc Commando was a popular large British parallel-twin sports touring machine.*

**1937 Panther 250cc.**
**Est. £1,000–1,200** *S*

**1960 Panther 100 594cc,** OHV twin port single cylinder engine.
**£1,500–2,000** *AT*

# GILERA *(Italian)*

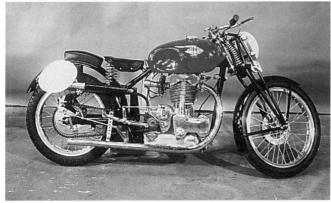

*l.* **1950 Gilera Saturno
Racing 500cc.
Est. £6,000–7,000** *S*

*Founded in 1909 Gilera was a major
force within the Italian motorcycle
industry until it hit financial trouble
and was taken over by Piaggio
(makers of the Vespa scooter) in 1969.
The 500 Saturno was to the Italians
what the BSA Gold Star was to the
British; being equally capable of many
tasks, including roadster, motocross,
trials and, as this example shows,
road racing.*

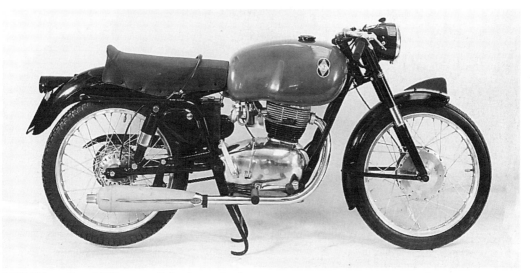

**1957 Gilera Sport 174cc,** parallel valve pushrod
engine, unit construction 4-speed gearbox, wet
sump lubrication.
**£1,000–1,500** *PC*

**1958 Gilera Saturno 500cc,**
finished in red and black.
**Est. £5,500–6,000** *BKS*

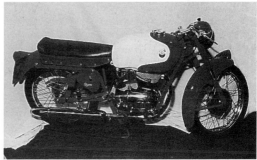

**1960 Gilera Extra 175cc.
£1,500–2,000** *IMO*

**1958 Gilera Extra 299cc,** OHV twin.
**Est. £1,550–1,650** *BKS*

**1975 Gilera 150 5V Arcore 150cc,**
fully restored condition.
**£750–1,250** *MR*

# GREEVES *(British)*

**1963 Greeves RAS Silverstone Solo Racing 250cc.**
£2,500–3,000 *S*

*In the early 1960s road racing in Britain was dominated at club and national level by the traditional big singles from AJS, Matchless and Norton. The 2-stroke was looked upon with disdain; many pundits openly voicing the belief that such machines were only suitable as commuter bikes, or at best dirt bikes. In 1962 Reg Everett road raced a Greeves with a tuned motocross engine. His success spurred the Thundersley, Essex, factory to build its first racers. These were powered by a 246cc Villiers engine, and labelled the RAS Silverstone.*

**1964 Greeves 25DC 250cc,** twin cylinder, fair condition, almost complete, finished in pale blue with polished mudguards.
£400–500 *S*

**c1965 Greeves Anglian Trials 250cc,** in good order, quite rare.
£1,500–1,800 *SW*

**c1965 Greeves Scottish Trials 250cc,** in good order.
£1,000–1,200 *SW*

**1964 Greeves Works ISDT Team Trials 246cc,** 2-stroke single cylinder engine, 4-speed gearbox, leading link front forks, good condition.
£3,000–4,000 *S*

**1964 Greeves Silverstone RBS Solo Racing 250cc,** non-standard Yamaha fairing and seat.
£1,500–2,000 *S*

*April 1964 saw the Silverstone RBS appear. This featured a Greeves top end with an Alpha crankshaft and Swedish Stefa flywheel magneto.*

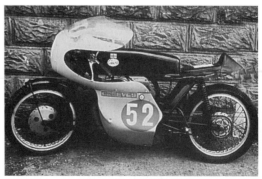

**1965 Greeves Silverstone RCS 250cc.**
£3,000–4,000 *S*

*Compared to the tiny RBS production figures, the 66 RCS models seems quite large. All were built during March and April 1965. This was the first production Greeves to feature a 2-leading shoe front brake.*

# HARLEY-DAVIDSON *(American 1903–)*

The United States can rightly be regarded as one of the true pioneering nations in the birth of the motorcycle, both for street and competition use. The first Harley was a 3hp side valve single, the Grey Ghost, built in 1903 when pattern maker Arthur Davidson teamed up with draughtsman William Harley to build their first bike, with Davidson's brother Walter, also lending a hand.

However, it was not until 1907, when working from little more than a small wooden hut, that Harley-Davidson was incorporated as a company and moved into more substantial premises. From then on progress was rapid in the extreme. By the time the USA entered WWI in 1917, Harley-Davidson was building an astonishing 17,000 motorcycles a year.

The company's first V-twin, a 800cc 45° model, had been launched in 1909 and started a trend which still exists today. During the war some 20,000 V-twins were supplied to the Allied forces, most with sidecars, many seeing military service in Europe.

Post-war main production centred on 1000 and 1200cc V-twins, but the Company suffered in decreased sales from an abundance of cheap cars, led by the Model T Ford.

In racing, Harley was soon challenging Indian for the top spot and, like their rivals, made successful trips to Europe in search of race victories. Joe Petrali was Harley's number one rider for much of the 1920s and 1930s. During the early 1940s Harley-Davidson built mainly military bikes once again.

AMF took over the company at the end of the 1960s and around the same time the XR750 racer made its bow. This found fame in the next decade thanks to men such as Rayborn, Brelsford and Springsteen, who raced both on dirt and tarmac.

A management buy-out occurred in the early 1980s and with it came the 'new nostalgia', in which Harley's V-twin street bikes were very much the bike to be seen on.

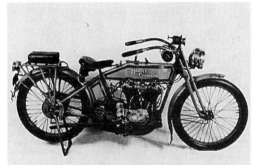

**1915 Harley-Davidson F,** standard 3-speed with high performance K engine, carbide lights, passenger rests, hand Klaxonet horn, original unrestored condition, complete, runs well.
**£37,500–42,500** *Gen*

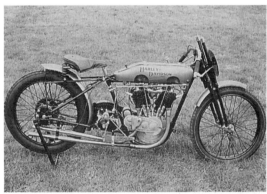

**c1918 Harley-Davidson T 1000cc,** V-twin engine, fully restored, very good condition, no modifications from original specification.
**£5,500–6,500** *BKS*

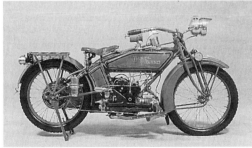

**1920 Harley-Davidson W 584cc,** opposed in-line, twin side valves, 4-stroke, bore and stroke 69 x 76.2mm, rear brake drum operated internally and externally by foot and hand.
**£8,500–9,500** *C(A)*

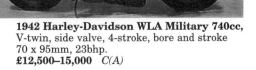

**1942 Harley-Davidson WLA Military 740cc,** V-twin, side valve, 4-stroke, bore and stroke 70 x 95mm, 23bhp.
**£12,500–15,000** *C(A)*

**1946 Harley-Davidson WRTT Racer,** engine recently overhauled, wide tanks, large aluminium oil tank.
**£10,000–11,000** *Gen*

**1948 Harley-Davidson WLC Civilian model** 750cc, 45° V-twin cylinder.
**£7,000–7,500** *BLM*

**1948 Harley-Davidson WR Racer,** recently overhauled engine, chrome moly flat track frame, spindle wheels, narrow tanks, dual sprocket rear wheel.
**£9,000–10,000** *Gen*

**1955 Harley-Davidson,** V-twin, major restoration, finished in bright red livery, whitewall tyres, black leather saddle, chrome plating.
**Est. £8,000–10,000** *S*

*l.* **1970 Harley-Davidson Electra Glide 1200cc,** wheels re-built, brakes overhauled, blue metallic finish.
**Est. £6,000–7,000** *LF*

**1971 Harley-Davidson FX Superglide 1200cc,** very good condition throughout.
**Est. £8,000–8,500** *BKS*

*l.* **1972 Harley-Davidson XR750 750cc.**
**£15,000–16,000** *S*

*This is the actual XR750 raced by Scott Brelsford, the brother of Mark who was AMA Grand National Champion in 1972.*

# HERCULES/DKW
## (German)

**1978 Harley-Davidson SXT 124cc,** single cylinder, 2 stroke, 5-speed, oil pump lubrication.
**£300–400** *S*

**1980 Hercules DKW Wankel 2000,** rotary engine, very tidy, rare and collectable machine.
**£2,200–2,500** *BLM*

# HONDA *(Japanese 1946–)*

Soichiro Honda was born in 1906, the eldest son of a village blacksmith in Komyo, long since swallowed up by the urban sprawl of modern-day Hamamatsu. He left school in 1922, taking up an apprenticeship in Tokyo, as a car mechanic. Later he returned to Hamamatsu, opened his own garage business and with his new found source of income went auto racing. This came to an abrupt end after a serious accident at the Tama River circuit, near Tokyo, in July 1936.

Following this he sold his dealership and entered the world of manufacturing with a piston ring factory. Then came the war, which saw the Honda company making aircraft propellers.

In 1946 Honda came back into the business world setting up the Honda Technical Research Institute. Unfortunately, this grand sounding venture was based in a tiny wooden hut, little more than a garden shed, on a levelled bomb site on the fringe of Hamamatsu.

For once luck was on his side and after uncovering a cache of 500 war surplus petrol engines, Honda launched himself on the path of motorcycle manufacture; something he was to do with unparalleled success.

The easy sale of these bikes encouraged him to move into motorcycle design. Shortly after the incorporation of Honda Motor Company in 1948, they produced over 3,500 98cc Model D 2-strokes and by 1953 produced 32,000 Model E 4-strokes.

Despite increasing production numbers Honda realised that to survive it would have to export bikes on a grand scale. To achieve this Honda used a combined strategy based on clever advertising, producing world championship winners, manufacturing smaller, efficient but reliable bikes at affordable prices. This recipe for success worked perfectly and by the mid-1960s Honda's production levels had reached 130,000 bikes per month.

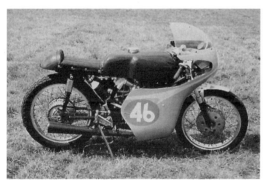

**1961 Honda/Hondis CB72 Racing,** road racing machine, right hand gear change to a close ratio 4-speed gearbox, twin leading shoe front brakes, remote float Amal carburettors with high lift camshaft and enlarged valves.
**Est. £2,800–3,000** *BKS*

**1963 Honda CR93 124cc.**
**£17,000–20,000** *S*

*The CR93 production racer is acknowledged as one of Honda's best ever designs. Generally it was not only a race winner, but also easy to ride. There was absolutely no power below 5,800rpm, but from then on to its peak at 13,000rpm, it was both powerful and vibration free. This example has the benefit of the works type double-sided, single-leading shoe front brakes.*

**1964 Honda C95 Benley 148cc,**
lightweight touring motorcycle.
**£1,000–1,500** *VMCC*

**1964 Honda CB92 Super Sport,** lightweight, 2 cylinder air-cooled unit, 4-speed gearbox, restored to concours condition.
**£4,500–5,000** *S*

1965 **Honda CB160,** lightweight twin, fully restored.
**£1,200–1,400** *S*

1965 **Honda CB160 161cc,** single camshaft, 360° crankshaft, electric starter.
**£900–1,000** *S*

1965 **Honda C200 89cc,** lightweight, restored, new wheel rims, stainless spokes, new tyres, speedometer, top end overhaul.
**Est. £1,200–1,300** *BKS*

1966 **Honda CB450 445cc,** restoration project.
**£220–320** *S*

*This example of Honda's 'Black Bomber' has been off the road in dry store for 21 years since 1973 and requires restoration.*

> **Miller's is a price GUIDE not a price LIST**

1967 **Honda CL77 305cc,** finished in silver, original condition.
**Est. £2,300–2,700** *S*

1967 **Honda CD175,** overhead cam twin cylinder engine, excellent mechanical and cosmetic condition.
**Est. £1,200–1,400** *S*

1968 **Honda CB750 USA 748cc,** 4-cylinder, OHC, 4-stroke, bore and stroke 62 x 62mm, 79bhp.
**£3,500–4,000** *C(A)*

1970 **Honda CB750 748cc,** 4-cylinder, air-cooled.
**£2,500–3,000** *BKS*

**1971 Honda CB750 748cc,** 4-cylinder, air-cooled, 5-speed gearbox.
**Est. £2,500–3,500** *BKS*

**1972 Honda CB750 748cc.**
**£1,200–1,300** *S*

*Apart from a fairing and racing-type saddle this machine is in original trim.*

**1976 Honda CB400F,** 400/4, 6-speeds, original tool kit.
**£1,300–1,700** *BKS*

**1978 Honda CBX1000 1047cc,** 6-cylinder, DOHC, 4-stroke, bore and stroke 64.5 x 53.4mm, 100bhp.
**£3,500–4,000** *C(A)*

# HRD *(British)*

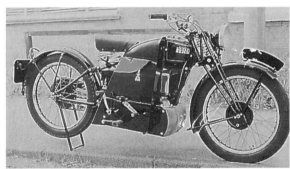

> Approximately 1,000 HRDs of all types were built in Wolverhampton, but less than 20 machines are thought to exist today.

*l.* **1934 HRD MOD W 499cc.**
**£7,000–7,500** *AtMC*

# HUMBER *(British)*

**1924 Humber SV 2¾hp.**
**£1,750–2,250** *AT*

*l.* **1928 Humber 349cc,** quite rare vintage OHV model, in good condition.
**Est. £3,800–4,200** *BKS*

**1928 Humber 350 349cc,** side valve.
**£3,250–3,750** *S*

*The Humber marque originated in the bicycle industry and was involved with 2, 3 and 4-wheel transport from the pioneer days. They enjoyed one moment of glory when P.J. Evans won the 1911 Junior TT, and some success at Brooklands, but for most of the 1920s were not dominant. In 1928 they introduced a series of 350cc engines and this machine has one of the latter. The production of all Humber motorcycles ceased in 1930.*

# INDIAN *(American 1901, British 1951–53)*

Indian, together with Harley-Davidson, is without doubt the most well-known of all American motorcycle marques. The Company was founded in 1901 by two former racing cyclists, George M. Hendee and Carl Oscar Hedstrom.

Indian's first production roadster, (it also built highly successful racing machines in these early days) was a 4-stroke single with vertical cylinder. With this design and the famous V-twin which first appeared in 1907, Indian soon developed a reputation for sophisticated design and excellent quality which was to stand for many decades.

One of the twins, a 600cc model was sent to Britain in the year of its launch and competed in the 1907 ACU Thousand Mile Trial. This event was later to become the famous ISDT (International Six Days Trial).

With a further eye to export across the Atlantic, Indian entered no less than four works riders in the 1911 Isle of Man Senior TT. Scoring an impressive 1-2-3, with the help of its newly created 2-speed gearboxes,

Indian's reputation in Europe was cemented and in the following year over 20,000 machines were exported. By 1919 the Scout, a 600cc side valve, was introduced and was soon followed by the Chief and the 1200cc Big Chief.

The introduction of mass produced cars in the USA, combined with import tariffs in the UK by the mid-1920s, followed by the Great Depression in 1929, hit Indian sales hard. However, it was still able to acquire the Ace marque but was eventually taken over in 1930 by E. P. Du Pont.

For the remainder of the 1930s and during WWII Indian soldiered on with profits still proving difficult to create. In 1949 a cash injection was made by British entrepreneur John Brockhouse. Brockhouse assumed control of Indian, but this failed to halt the company's financial slide and production was terminated in 1953. Since then many have tried to relaunch the name, including American publisher Floyd Clymer, but without success.

**1908 Indian Camel Back,** twin HP3, original condition.
**£30,000–40,000** *IMC*

*This is reputed to be the oldest Indian motorcycle in the UK.*

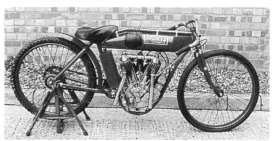

**c1910 Indian Model D 7hp,** V-twin, single speed countershaft gear, Splitdorf magneto ignition, mechanically good, frame and forks good, paintwork excellent, original saddle, tank needs repair.
**Est. £7,000–7,500** *BKS*

**1914 Indian Boardtrack Racing 1000cc,** auxiliary pedal starting, overhead inlet valve engine.
**£8,500–9,000** *BKS*

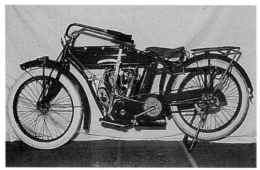

**1913 Indian Twin 7hp,** partially restored to original specification.
**£15,000–16,000** *IMC*

*r.* **1922 Indian Scout 600cc,** V-twin, twin rear brakes, operated internally by foot, externally by handle, left foot clutch, left hand throttle, right hand advance.
**£7,000–10,000** *ST*

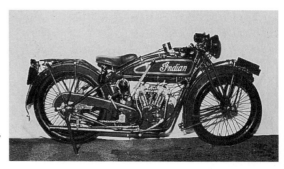

**1936 Indian Chief 1200cc,** restored over 6 years to original specification, finished in mohawk green/seminole cream.
**£16,000–18,000** *IMC*

**1929 Indian Scout 750cc,** original condition.
**£8,000–10,000** *IMC*

**1940 Indian Chief 1200cc,** 74C engine, restored to original condition.
**£12,000–18,000** *IMC*

**1940 Indian 648 Racer,** good condition.
**£12,000–13,000** *Gen*

*This model is fitted with one of only 50 engines made specially by the company.*

**1942 Indian 741A Scout 493cc,** V-twin, side valve, 4-stroke, bore and stroke 63.4 x 77mm, 8hp.
**£9,000–10,000** *C(A)*

*This was produced as a military model and has been totally restored for civilian use.*

**c1942 Indian Military 741,** engine recently overhauled, leather saddle bags, lighting, generator missing.
**£4,000–5,000** *Gen*

**1947 Indian 741B Civil Trim,** V-twin.
**£4,500–5,000** *BLM*

**1951 Indian Road Warrior 500cc,** V-twin, Western handlebars, original except repainted.
**£4,500–5,000** *Gen*

*l.* **1953 Indian Chief 1340cc,** 80cu in engine, mostly original, partly restored.
**£18,000–20,000** *IMC*

*The above price would apply if the machine was restored for road use.*

# IVY (British)

**1925 Ivy 300cc.**
**£5,000–6,000**  *AtMC*

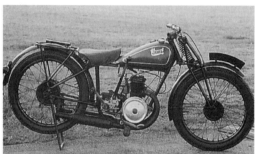

**1929 James Sports 172cc,** twin-port Villiers powered, finished in black with metallic brown fuel tank, good condition.
**£1,200–1,500**  *S*

**1949 James Comet 98cc.**
**£400–600**  *BLM*

**1957 James Captain 197cc.**
**£500–600**  *BLM*

# JAMES (British)

**1925 James Flat Tanker 3hp.**
**£3,000–3,500**  *BLM*
*This is a rare model.*

**1947 James Comet 98cc,** restored.
**£650–750**  *PC*

**1966 James Sports Superswift 249cc,** 2-stroke.
**£1,000–1,800**  *PC*

# JAWA (Czechoslovakian)

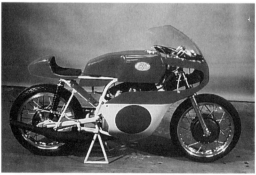

**1967 Ex-works JAWA 350 GP Racing,**
water-cooled 2-stroke, in good running order.
**Est. £8,000–9,000**  *S*

*This particular machine is one of four built as a single to keep them competitive. It is eligible for both international historic racing and British classic racing in which it would be extremely competitive.*

# KAWASAKI *(Japanese)*

**1970 Kawasaki F5 Big Horn 350cc,** single cylinder engine, 4-speed gearbox, imported from USA in 1978, good condition.
**Est. £2,000–2,200** *BKS*

**1975 Kawasaki Z1 903cc,** 4-cylinder DOHC, good condition, all original except exhaust, now 4:1, originally 4:4.
**£2,500–3,500** *PC*

*Imported from the USA in 1991.*

**1975 Kawasaki Z1 903cc,** transverse 4-cylinder, 4-stroke, good condition.
**£4,500–5,000** *S*

**1975 Kawasaki Z1 903cc,** 4-cylinder, in-line DOHC, bore and stroke 66 x 66mm, 81bhp
**£4,200–5,200** *C(A)*

*Honda produced the first of the 'Big Fours' to emerge from Japan with a capacity of 750cc. Kawasaki then produced the larger capacity Z1, which, with its acceleration became an immediate best seller.*

# LAVERDA *(Italian 1948–)*

The very first Laverda appeared in 1948, after it had taken its builder Francesco Laverda some 12 months to design and construct the prototype. Initially the concept was not for production, but his own personal use. It was only later, when a number of his friends in the north-eastern town of Breganze saw the machine that their enthusiasm led to the start of a famous lineage. This first model had a 75cc OHV single cylinder engine. Descendants of this model continued throughout the 1950s. In the 1960s came a small 4-stroke scooter (also built under license by Montesa) and the company's first twin. This latter model was powered by a 200cc 4-stroke unit construction engine but was more touring than sport.

The annual show at Earls Court, London was chosen for the launch of the Laverda 650 twin. This was to lead to one of the marque's most famous models, the 750 SFC. Altogether nearly 18,500 big twins were manufactured over the next decade or so.

A prototype of the 3-cylinder Laverda appeared in 1970 and was to herald a decade of success for the Breganze marque, culminating in the 140mph Jota sportster.

Unfortunately Laverda also made some costly mistakes, including failures such as

the Zündapp powered 125/175cc 2-stroke roadsters and 125/250 enduro bikes – again powered by 2-strokes motors.

One of the sensations of the 1981 Milan Show was the new RGS, but even this bike, using one of the legendary 1000cc triple engines, failed to attract customers in sufficient numbers to stave off a financial crisis. This came to a head in the late-1980s after several owners and a series of promises did not materialise, and Laverda was eventually relaunched in the early 1990s with a single model range, the excellent 650 (668cc) Sport DOHC twin.

**1971 Laverda 750SF,** twin cylinder, drum brakes.
**£2,500–2,800** *PC*

# LEVIS *(British)*

**1924 Levis T2 247cc,** 2-stroke, single cylinder, belt drive, 2-speed gearbox and kickstarter.
**£1,700–2,000** *PS*

**1930 Levis A30 346cc.**
**Est. £2,000–2,500** *HOLL*

*Besides its well-known 2-strokes, the Levis concern also marketed a range of 4-stroke models. These had vertical cylinders, rear magneto and total loss lubrication.*

**1934 Levis A34 346cc.**
**Est. £1,800–2,500** *HOLL*

*In 1934 the A-series became the A34; there were options of engine tuning parts, an improved exhaust system, and a new frame with a 4-speed option.*

**1935 Levis A35 350cc,** OHV.
**£1,400–1,700** *HOLL*

**1939 Levis SF 350cc,** for restoration.
**£400–600** *HOLL*

**1936 Levis Special D 498cc.**
**Est. £1,000–1,500** *HOLL*

# MAGNAT-DEBON *(French)*

*r.* **c1924 Magnat-Debon 250cc,** air-cooled inclined single cylinder engine, magneto ignition, Gurtner carburettor, 2-speed hand change gearbox, electric dynamo lighting, speedometer, luggage carrier and tool box, to original specification.
**Est. £800–1,000** *BKS*

# MATCHLESS *(British 1901–69, revived 1987)*

Harry Collier and his sons Charlie and Harry Junior were true pioneers of the British motorcycle industry. They began their activities in London in 1901 using De Dion and MMC engines initially, having built an experimental Matchless back in 1899.

Not only did the Collier family build the bikes, they also rode them with great success. Harry Collier Junior won the 1903 ACU 1,000 – Miles Reliability Trial and Charlie Collier won the Single Cylinder Class of the first Isle of Man TT race in 1907. Two years later Harry Collier won the TT at a record average speed of 49mph, whilst the following year the two brothers finish first and second in the TT – a feat unsurpassed today.

During the Great War production switched to aircraft parts and rifle bayonets. However, post-war saw a return to two wheels – plus sidecars – and even the Model K car (in 1923). In 1928 the firm officially became Matchless Motor Cycles Ltd. Then in 1930 the famous Silver Arrow V-twin was produced, followed by even more glamorous

Silver Hawk V4 the following year. That same year, 1931, Matchless took over AJS. More expansion followed in 1937, with the acquisition of the Sunbeam marque, and the 'group' being re-registered under the Associated Motor Cycles (AMC) banner.

During WWII AMC were a major supplier of military motorcycles to the British armed forces, with the G3L (Lightweight) proving a firm favourite with many wartime despatch riders, using AMC's version of a BMW hydraulically damped telescopic front fork. In the immediate post-war period the bulk of AMC's production was exported. In 1951 the famous 'jam pot' rear suspension units were introduced.

Then came a period of further expansion with the purchase of Francis-Barnett, James and Norton. However, this proved to be AMC's downfall. First, the decision to manufacture two-stroke engines for James and Francis-Barnett, second the mix of Norton and AMC parts to create special bikes for the USA.

**1925 Matchless 500cc.**
**£5,500–6,000** *AtMC*

**1930 Matchless Silver Arrow Model A 394cc,**
narrow angle V-twin, side valve, common camshaft, 4-stroke, cylinders in one block.
**£5,000–6,000** *C(A)*

**1939 Matchless Model X 1000cc,** V-twin.
**£5,500–6,500** *BLM*

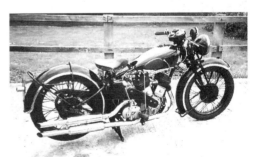

**1936 Matchless Model X 1000cc.**
**£5,000–6,000** *BLM*
*This model, with its longer frame, can adequately take a sidecar.*

*l.* **1943 Matchless G3L 350cc.**
**£900–1,200** *S*

*A familiar sight during WWII, the Matchless 350 G3L was probably Britain's best military motorcycle of its period. This was due to the combination of an excellent overhead valve engine and the luxury of telescopic front forks.*

**1946 Matchless G80 500cc.**
**£2,500–3,000** *BLM*

**1948 Matchless G3 350cc.**
**£1,000–2,200** *BLM*

**1949 Matchless G3L 350cc.**
**£1,200–1,500** *BKS*

**1952 Matchless G9 Spring Twin.**
**£2,500–3,000** *BLM*

**1953 Matchless G3LS in AFS Livery.**
**£1,850–1,950** *VER*

**1953 Matchless G3LS 347cc,** original, in good running order.
**Est. £1,200–1,800** *LF*

**1955 Matchless G80S**
**Competition Trials 500cc.**
**£3,000–3,500** *BLM*

**1955 Matchless G3LS 350cc,** restored, with period extras including fairing and panniers.
**£1,300–2,500** *NAC*

*l.* **1956 Matchless G9 Super**
**Clubman 500cc,** fully restored.
**Est. £3,000–3,200** *BKS*

**1956 Matchless G80S 498cc,** single cylinder, OHV, 4-stroke, bore and stroke 82.5 x 93mm, 23.5bhp.
**£2,500–3,000** *C(A)*

**1957 Matchless G11 'CSR Copy' 650cc.**
**£2,500–3,000** *BLM*

**1957 Matchless G3L 350cc,** overhead valve single cylinder engine, plunger rear suspension, telescopic forks, separate gearbox, good condition.
**Est. £1,250–1,400** *S*

**c1956 Matchless Competition G3,** unusual modified machine, frame top tube replaced by a stressed member aluminium fuel tank, allowing exceptional lock, wheelbase shortened at the swinging arm.
**£500–600** *S*

**1959 Matchless G12 650cc.**
**£3,000–3,300** *BLM*

**1958 Matchless G11 600cc,** twin.
**£1,500–1,800** *HOLL*

**1959 Matchless Trials G3LC 350cc.**
**£3,500–4,000** *BLM*

**1960 Matchless Metisse Motocross G80CS 500cc,** in original condition.
**£2,500–3,000** *SW*

*l.* **1960 Matchless G50 Racer 496cc,** single cylinder, OHC, 4-stroke, bore and stroke 90 x 78mm, 49 bhp.
**£14,000–16,000** *C(A)*

*This racing machine was a derivative of the AJS 7R 'Boy Racer' and became the great rival of the Manx Norton. AJS and Matchless were both being produced by Associated Motor Cycles of London.*

**1961 Matchless G5 350cc.**
**£2,500–3,000**  *PS*

*This model was never as popular or respected as the heavyweight brother, the lightweight AMC single, sold under the Matchless (and AJS) labels which featured pushrod unit construction engines in both 250 and 350cc.*

**Pre-1965 Matchless Metisse Scrambler 500cc.**
**£1,000–2,000**  *PS*

*The brothers Don and Derek Rickman of New Milton, Hampshire, did much to boost British prestige with their Matchless Metisse Scramblers in the 1960s. These machines are now much sought after for pre-1965 Scrambler events.*

*l.* **1961 Matchless G12 650cc.**
**£1,800–2,000**  *BLM*
*This machine is in period café racer trim.*

# METRO-TYLER *(British)*

**1923 Metro-Tyler 269cc,** single speed belt drive, in good condition.
**£1,400–1,800**  *S*

*The Metro-Tyler started life as the Metro in Saltley, Birmingham, in 1912, but post-WWI the firm were taken over by the Tyler Apparatus Company Ltd of Kilburn Lane, London, in 1919 This example is the 2½hp 'All Black Baby', believed by the VMCC marque specialist to be the last survivor of the model. It was so-called because they were painted all black, rather than the Indian red of the earlier models.*

# McKENZIE *(British)*

**1913 McKenzie Ladies 169cc,**
2-stroke, open frame.
**£1,000–1,500**  *S*

# MOTOBECANE *(French)*

**c1928 Motobécane Solo,** 2-stroke single cylinder, 2-speed gearbox, kick start, chain-cum-belt drive, restored to original specification.
**£700–800**  *BKS*

# MOTOBI *(Italian)*

**c1960s Motobi Arrow 125cc.**
**£600–1,200**  *NLM*

# MOTO GUZZI *(Italian 1921–)*

Moto Guzzi's progress is a journey through motorcycle history. From only ten employees in 1921, including the two partners, Carlo Guzzi and Giorgio Parodi, to Italy's largest and most famous factory employing hundreds, from glorious success, to the verge of bankruptcy, new ownership brought success again into the 1970s and 1980s.

From the firm's entry into racing in the early 1920s until its withdrawal in 1957, it favoured a horizontal single cylinder engine layout, and with this basic design Guzzi machines and riders won ten TTs and eight World Championships.

Although Moto Guzzi achieved numerous Continental victories in its early days it was not until 1935, when Stanley Woods won both the Lightweight and Senior TTs, that Guzzi's efforts were crowned with truly International success. Other notable pre-war victories came in the 1937 Lightweight TT and the epic

defeat of the mighty DKWs in the 1939 250cc German Grand Prix.

Remaining faithful to the horizontal single cylinder, Guzzi largely dominated the post-war 250cc class from 1947 until 1953. Then the Mandello del Lario company enlarged the engine capacity, entered the 350cc class and astounded the racing fraternity by winning the world title at its first attempt.

Guzzi will always be remembered for its amazing versatility in design, for in addition to the famous singles it produced machines with V-twin, across-the-frame 3-cylinders, in-line 4-cylinders and even a V8 engine!

After financial decline in the 1960s Guzzi was reborn in the 1970s with the backing of the Argentinian industrialist De Tomaso and the success of its range of V-twin touring and sports bikes. Descendants of these bikes still form the backbone of Guzzi's current range of motorcycles.

**1937 Moto Guzzi Egretta Single 250cc,** fully sprung frame, friction type shock absorbers, original, good running order.
**Est. £1,500–2,000** *BKS*

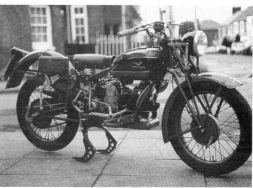

**c1939 Moto Guzzi ALCE 498cc,** OHV.
**£4,000–4,500** *AtMC*

**c1939 Moto Guzzi ALCE 500cc,** horizontal single cylinder engine, the magneto rebuilt.
**Est. £4,000–4,500** *S*

**1951 Moto Guzzi Sport Falcone 498cc,** single cylinder, OHV, 4-stroke, 23bhp, recently restored.
**£4,200–5,000** *C(A)*

**1952 Moto Guzzi Airone Sport 247cc,** flat single cylinder, air-cooled, overhead valve, 4-stroke, 4-speed, external flywheel, front telescopic fork, rear friction damped swinging arm.
**Est. £2,200–2,400** *C*

**c1953 Moto Guzzi Moto Leggera 65cc,** rotary valve 2-stroke.
**£1,200–1,500** *AtMC*

**c1953 Moto Guzzi Airone 247cc,** horizontal
single cylinder, air-cooled, 4-stroke, with exposed
hairpin valve springs, red livery, running well and
in original condition.
**Est. £2,000–2,200** *BKS*

**1956 Moto Guzzi Falcone 500cc,** outside
flywheel, horizontal single cylinder engine,
good condition.
**Est. £4,200–5,200** *BKS*

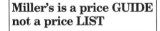

Miller's is a price GUIDE
not a price LIST

**1968 Moto Guzzi V7 700cc,** transverse 90°
V-twin cylinder, overhead valve, 4-stroke, air-
cooled, 5-speed, front telescopic fork, rear
swinging arm, shaft drive.
**Est. £2,500–2,700** *C*

**1953 Moto Guzzi Falcone Sport 500cc.**
**Est. £4,200–5,200** *BKS*
*The Falcone Sport had a top speed of 85mph and
developed 23bhp at 4,300rpm.*

**1955 Moto Guzzi Falcone 500cc,** restored to
concours condition including engine, gearbox,
brakes and suspension rebuild.
**£4,200–5,200** *BKS*

**1967 Moto Guzzi V7 703cc,** good condition.
**£1,000–1,500** *BKS*

**1974 Moto Guzzi V7 Sport 748cc USA model,** transverse 90° V-twin cylinder, overhead valve, 4-stroke, air-cooled, 5-speed, front twin discs, rear drum, front telescopic fork, rear swinging arm, shaft drive.
**Est. £3,400–3,600** *C*

**1980 Moto Guzzi California 844cc,** king and queen seat and Dunstall silencers.
**£2,000–2,500** *IMO*

# MOTO MORINI *(Italian)*

**1963 Moto Morini 175F3 174cc,** OHV Formula 3 racer, close ratio gearbox.
**£8,000–8,500** *AtMC*

**1977 Moto Guzzi Le Mans MK1 844cc,** twin cylinder engine, 5-speed gearbox, good mechanical condition, paintwork excellent.
**Est. £3,000–3,200** *BKS*

**c1973 Moto Morini 3½ Strada.**
**£700–1,500** *NLM*

*Early models had wire wheels, TLS brake and chrome rims.*

**c1977 Moto Morini 3½ Sport,** mag wheels, disc brakes, with an optional fairing.
**£1,500–2,500** *NLM*

**1975 Moto Morini 3½ Sport 344cc,** V-twin, OHV,
4-stroke, 35bhp.
**£2,500–3,000**   *C(A)*

*A native of the university town of Bologna, Alfonso
Morini was born in 1892. In the early 1920s he
helped found the MM marque, leaving them to form
Moto Morini in 1937. During the immediate post-
war period Morini built a number of small single
cylinder models with both 2 and 4-stroke engines.
Then in the early 1970s the company built the 3½
Strada. This featured Heron cylinder heads and a
Sport model as seen here was soon added.*

**1975 Moto Morini 3½ Strada 350cc.**
**£700–1,500**   *S*

*The early Strada models had wire wheels and drum
brakes. In 1975 Morini updated the machine with
the addition of cast alloy wheels and a Grimeca
hydraulically operated front disc. The engine was a
72° V-twin of 344cc which, driving through a
6-speed gearbox, gave the machine a maximum
speed of just under 100mph. The Sport version was
soon 5mph faster.*

**c1977 Moto Morini 3½ Sport,** body kit and
2:1 exhaust system.
**£2,500–3,000**   *NLM*

*This model is a Valentini Special.*

> **Locate the source**
> *The source of each illustration in
> Miller's Classic Motorcycle Price
> Guide can easily be found by check-
> ing the code letters at the end of each
> caption with the Key to Illustrations
> located at the front of the book.*

# MUNCH *(German)*

# MOTO RUMI *(Italian)*

**1957 Moto Rumi 124cc.**
**£2,200–2,800**   *PCC*

**1971 Münch Mammoth 1177cc,** NSU Prinz
air-cooled 4-cylinder engine, produced in limited
numbers, original specification.
**£10,000–10,500**   *BKS*

**1961 Moto Rumi Junior Gentleman 124cc,**
horizontal twin cylinder engine.
**£2,200–2,800**   *IMO*

**1975 Münch 1200 TTS 1177cc,** 4-cylinder, in-line
OHC, 4-stroke, 88bhp.
**£9,000–10,000**   *C(A)*

# MV AGUSTA *(Italian 1945–1978)*

No other marque has more charisma than MV (Meccanica Verghera) Agusta. The legendary Count Domenico Agusta masterminded a race team which dominated much of the road racing world championships from the early 1950s through to the mid-1970s; with machinery ranging from single cylinder 125s to fire-breathing 4-cylinder 500s. There was even a prototype 6-cylinder model.

During the 28 years it contested the sport (from 1948–1976) MV won over 3,000 races and 37 world championships. The list of riders who rode for the marque reads like a *Who's Who* of motorcycle sport, with Sandford, Graham, Lomas, Ubbiali, Provini, Surtees, Taveri, Hocking, Hailwood, Agostini and Read prominent among MV's leading stars. MV's achievements were all the more amazing because, unlike the majority of other famous Italian racing factories, it had no pre-war competition history. In fact its first ever bike (a humble 98cc two-stroke commuter) did not appear in prototype form until 1945!

In stark contrast to its racetrack successes, MV's production roadsters were largely non-inspirational. Initially focusing more on scooters rather than motorcycles, attempts to export their products, even the 4-cylinder sportsters, became a fruitless effort. In the 1970s the helicopter business had grown in importance, and as a result the racing and road-going bikes ceased production. Even so its products have steadily risen in esteem to become a cult status within the classic bike scene, even its most basic bikes attract both the collector and enthusiast alike.

**1949 MV Agusta Competizione 125.**
**£7,000–8,000** *BKS*

*Only about 100 of these 2-stroke machines were originally constructed.*

**1956 MV Agusta 175 OHV Disco Volante,** red livery, rebuilt and restored.
**Est. £3,000–3,400** *BKS*

**1956 MV Agusta Sport Turismo Rapido 125,** OHV, rebuilt with many new parts, red livery.
**Est. £2,000–2,200** *BKS*

*This bike has the signatures of John Surtees and Giacomo Agostini on the petrol tank.*

**1957 MV Agusta CSS 172.4cc,** rare double overhead camshaft model, original condition.
**£21,000–22,000** *MVA*

*One of only 12 machines built and the legendary Mike Hailwood raced one early in his career.*

**1971 MV Agusta 125 Sport,** OHV unit construction engine, 4-speed gearbox.
**Est. £1,200–1,500** *BKS*

**1972 MV Agusta 750S 743cc,** 4-cylinder, DOHC, 4-stroke, 78bhp.
**£16,000–18,000** *C(A)*

**1973 MV Agusta 750S,** overhead camshaft engine, 5-speed gearbox, electrics, paintwork and machine in good condition.
**£16,500–17,500** *S*

**1973 MV Agusta 750S 743cc,** totally restored to original condition.
**£16,000–18,000** *MVA*

**c1973 MV Agusta 350B 349cc,** twin-cylinder racer, red livery, good working order.
**£1,800–2,000** *BKS*

**1974 MV Agusta 750S,** 4-cylinder.
**£16,000–18,000** *PCC*

**1976 MV Agusta 750S America 789.7cc,** 4-cylinder, DOHC, 4-stroke, 75bhp at 8500rpm.
**£15,000–17,000** *C(A)*

**1975 MV Agusta 125 Sport 124cc,** 5-speed gearbox, rare, excellent condition.
**£3,000–3,500** *S*

**1975 MV Agusta 350 Sport,** Ceriani 32mm front forks, cast alloy wheels, in good running order.
**Est. £2,500–2,800** *BKS*

**1975 MV Agusta Mini 48/10 47.6cc,** single cylinder 2-stroke engine.
**£1,500–1,600** *PC*

**1976 MV Agusta Mini 48/10 47.6cc,** single cylinder 2-stroke engine, 2 bhp.
**£14,00–16,000** *C(A)*

**1976 MV Agusta 125 Sport 124cc.**
**£2,800–3,000** *S*

*Specification for this miniature MV Agusta includes a bore and stroke of 53 x 56mm, compression ratio 9.8:1 and 14bhp at 8500rpm. The factory claimed 75mph. The gearbox was a 5-speed. This particular machine was one specially imported by the former Ducati and Cagiva importers company for Mick Walker's personal use inscribed 'Mick Walker, Wisbech, Cambs.' logos on the side panels.*

**1977 MV Agusta America 750cc.**
**£13,000–14,000** *PCC*

**1977 MV Agusta 350 Sport 349cc.**
**£2,500–3,500** *S*

*With triple disc brakes, 'spider web' cast alloy wheels, superb handling and a 105mph maximum speed the 350 Sport had a lot going for it. It was a shock to learn the valves were pushrod operated and the engine suffered quite serious vibration.*

**1977 MV Agusta 350 Sport 349cc,** twin cylinder.
**£3,800–4,200** *S*

*This example of MV Agusta's Sport 350 twin has the optional fairing, which certainly adds considerable grace to the machine. This cost an extra £90.72 in 1977.*

# NER-A-CAR
*(American / British)*

**1921 Ner-a-Car 2¼hp.**
**£4,000–5,000** *S*

**1978 MV Agusta 350 Sport Ipotesi Model 216.**
**£3,000–3,500** *MR*

# NEW HUDSON (British)

**1915 New Hudson C Tourist 2¼hp 211cc,**
2-stroke single cylinder engine, 2-speed
countershaft gear, extensive restoration.
**£700–900** *BKS*

**1957 New Hudson Re-Styled Autocycle 98cc.**
**£200–400** *NAC*
*This machine is known as the 're-styled' model. It
failed to compete and after a couple of years New
Hudson ceased production.*

# NEW IMPERIAL (British)

**1930s New Imperial 150cc,** OHV, single port.
**£800–1,000** *BLM*

**1935 New Imperial 250cc,** single hand change, in
original unrestored condition.
**£700–1,000** *BCB*

> ## Locate the source
> *The source of each
> illustration in* Miller's
> Classic Motorcycle Price
> Guide *can easily be
> found by checking the
> code letters at the end of
> each caption with the
> Key to Illustrations
> located at the front of
> the book.*

*l.* **1937 New Imperial 247cc,** single
cylinder, OHV, 4-stroke, 10 bhp.
**£3,500–4,000** *C(A)*

# NIMBUS (Danish)

**1945 Nimbus 750cc,** 4-cylinder, in-line, shaft
driven, rebuilt c1979, good condition.
**£2,200–2,900** *VER*

**1949 Nimbus Four 700cc,** in-
line, 4-cylinder, shaft drive.
**£1,600–2,000** *BLM*

# NORTON *(British 1902–)*

James Lansdowne Norton (often more simply referred to as 'Pa') built his first motorcycle in 1902, using a French Clement engine. This was soon complemented by another French engine, a Peugeot V-twin. It was one of these latter engines with which Rem Fowler won the Twin Cylinder race at the first TT in 1907. In the same year Norton designed the first engine of his own manufacture. This long-stroke engine, with 79 x 100mm, 490cc, dimensions, was to become something of a Norton tradition.

In 1911, after a long illness, 'Pa' lost control of Norton to the Vandervell family though he remained joint managing director with R. T. Shelley. He died in 1925 at the age of 56.

For much of the 1930s Norton was a major force on the racing scene with riders such as Jimmy Guthrie, Alex Bennett, Freddie Frith and Harold Daniel beating the world and winning no less than 14 TTs. Besides the riders, Irishman Joe Craig, a Norton rider himself during the 1920s, was largely the driving force in a race managerial career which was to span almost 30 years, from

1926–1955. Of course WWII stood in the way and Norton became a major supplier to the British Army during the conflict.

In the late 1940s and early 1950s Norton continued its racing successes with new riders, including Artie Bell, Geoff Duke and Ray Amm. Unfortunately this did not reflect in the sales of its roadsters success. A combination of slow sales and anticipation of death duties by the Vandervell family resulted in Norton being sold to AMC in 1952. In 1962 the company moved from its home in Bracebridge Street, Birmingham, to the AMC works in Plumstead. When AMC collapsed, new owners Manganese Bronze Holdings created the Commando using the Atlas machine in a new frame, featuring the 'Isolastic' system, designed to keep vibration to a minimum. After the Commando came a period of stagnation, before limited production of the Rotary began. Even racing successes in the late 1980s and early 1990s did not help and hence the future of Norton is in serious doubt.

**c1916 Norton 16H 500cc.**
**£3,000–3,200** *BLM*

**1921 Norton Brooklands Racing Special 490cc.**
**£8,000–9,000** *VMCC*

*Every BRS engine was tested at Brooklands and issued with a certificate to certify that it had exceeded a 70mph lap.*

**1925 Norton 16C Colonial 499cc,** 4-speed Sturmey-Archer gearbox controlled through a hand quadrant, girder forks with coil spring suspension and rigid frame at rear, excellent condition.
**£4,000–5,000** *S*

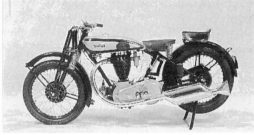

**1927 Norton Model 19 588cc,** single cylinder, OHV, 4-stroke.
**£4,000–5,000** *C(A)*

*The Model 19 was added to the Norton range in 1925 and continued unchanged until 1933, when the capacity increased to 597cc. In this latter form the Model 19 continued until the commencement of WWII, indicating the popularity of this reliable model.*

*l.* **1928 Norton Model 18 500cc.**
**£6,500–7,500** *BLM*

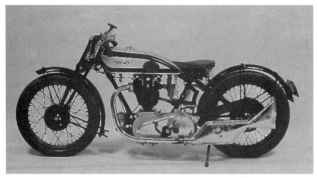

*l.* **1928 Norton Model 18 490cc,** single cylinder, OHV, 4-stroke, 7bhp.
**£4,000–5,000** *C(A)*

*The Model 18 made its debut at the 1922 Olympia Show and the following year broke the 12 hour record at Brooklands, amassing 18 world records in the process.*

**Use the Index!**

*Because certain items might fit easily into any number of categories, the quickest and surest method of locating any entry is by reference to the index at the back of the book.*

*This index has been fully cross-referenced for absolute simplicity.*

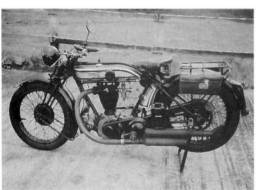

**1928 Norton Model 19 600cc,** exposed valve OHV flat tank.
**£6,500–7,500** *BLM*

**1934 Norton Bronze Head 350cc,** totally original.
**£9,000–10,000** *SW*

*An ex-works machine with documented history.*

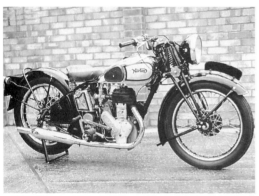

**c1937 Norton Big 4 633cc,** side valve, big single.
**£3,000–3,500** *AtMC*

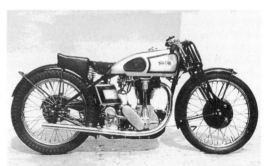

**1937 Norton International 490cc,** in race trim.
**£5,500–6,500** *VER*

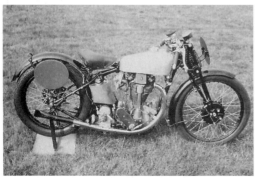

**1937 Norton International/Manx 499cc,** single cylinder overhead camshaft engine, rebuilt, 4-speed gearbox in good condition.
**Est. £5,500–6,000** *BKS*

**1937 Norton International Model 30 490cc,** single cylinder, OHC, 4-stroke.
**£6,500–7,500** *C(A)*

**c1938 Norton International 490cc.**
**£8,000–9,000** *AtMC*

**1939 Norton Model 16H 500cc.**
**£2,000–2,500** *BLM*

**1939 Norton Model 18 500cc.**
**£2,000–3,000** *BLM*

**1939 Norton Manx 499cc.**
**£7,000–7,500** *PM*

*l.* **1942 Norton Model 16H 500cc,**
side valve army despatch bike.
**£1,500–1,800** *BMM*

**1946 Norton Manx 499cc,** double overhead camshaft, 1939 Inter crankcase, engine rebuilt, race trim, in good running order, concours condition.
**Est. £10,500–12,000** *S*

**1948 Norton ES2 490cc,** OHV single.
**£1,500–2,500** *BLM*

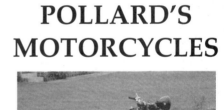

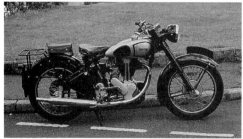

**1948 Norton ES2 490cc.**
£1,500–2,500 *PM*

**1949 Norton Big 4 597cc,** side
valve, single cylinder, rebuilt.
£1,400–1,750 *BCB*

**1950 Norton International Model 40M 348cc,**
single cylinder, OHC, 4-stroke, 24 bhp.
£8,000–8,500 *C(A)*

**1950 Norton International Model 30M 490cc.**
concours condition.
£8,000–9,500 *SW*

**c1948 Norton Manx 30M Beart Replica 490cc,**
overhead camshaft, painted in light green.
**Est. £10,000–10,500** *BKS*

**1950 Norton International Model 30M 490cc,**
single cylinder, OHC engine, road trim.
**Est. £8,000–8,500** *S*

**1950 Norton ES2 490cc,** single
cylinder, OHV, 4-stroke, 21bhp.
£1,000–2,500 *C(A)*

**1950 Norton 500T 490cc,** OHV.
£3,000–3,250 *PCC*

*l.* **1951 Norton 5T Trials 490cc,**
original specification.
**Est. £3,000–3,500** *S*

**c1952 Norton Manx 30 499cc,** DOHC, 4-speed gearbox, rebuilt.
**Est. £12,500–14,000** *BKS*

**1952 Norton ES2 490cc,** single vertical cylinder, air-cooled, overhead valve, 4-stroke, 4-speed separate unit, telescopic front fork, very good overall condition.
**£1,500– 2,500** *C*

**1954 Norton International Special 490cc,** mechanically in very good condition.
**£4,500–5,000** *S*

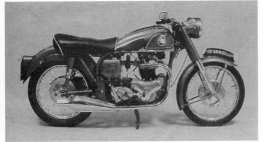

**1956 Norton 99 Dominator 596cc** vertical twin, OHV, 4-stroke, 31bhp.
**£3,000–4,000** *C(A)*

**1952 Norton ES2 490cc,** single, plunger rear suspension, black and silver, in near original condition.
**£1,250–2,250** *MR*

**1952 Norton 7 Dominator 496cc,** OHV vertical twin, completely rebuilt.
**£3,000–3,200** *PS*

**1952 Norton ES2 500cc,** single cylinder, near original condition.
**£1,000–1,700** *MR*

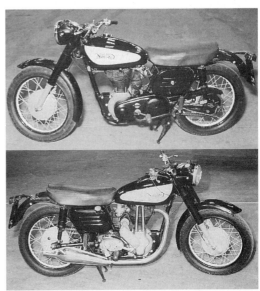

**1956 Norton ES2 490cc.**
**£1,500–2,000** *PS*

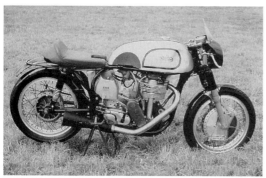

**1957 Norton Manx Replica 499cc.**
Est. £7,800–8,200   RKS

**1957 Norton Model 77 600cc.**
£2,700–3,100   BLM

*This is the Dominator in a single down tube frame.*

**1957 Norton Model 50 350cc.**
£2,250–2,750   PCC

**1957 Norton International Model 30 Featherbed 490cc,** single cylinder, OHC, 4-stroke, 29.5bhp.
£8,000–9,000   C(A)

*This model was the last 'Inter' built at the Birmingham Bracebridge Street works. Featuring the famous 'Featherbed' frame, which gave excellent road holding, it has now become one of the most sought after classic motorcycles.*

**1959 Norton Dominator 99,** full width hubs and wide line frame.
£2,600–3,300   BLM

**1959 Norton Dominator 99 600cc,** TLS front brake, Commando primary cases and inclined engine.
£2,000–2,400   BLM

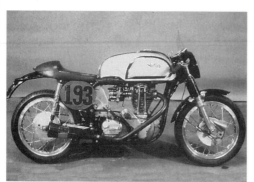

**1959 Norton ES2 490cc,** Quaife 5-speed gearbox, converted for classic racing.
Est. £3,000–3,500   S

**1959 Norton 99S 600cc,** standard specification, except alloy wheel rims, completely restored.
£3,000–3,500   NOC

**1960 Norton Model 50 350cc,** overhead valve single cylinder engine, to maker's specification and painted in traditional Norton livery of black and silver.
**Est. £1,900–2,200** *BKS*

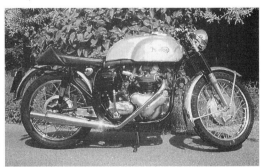

**1960 Norton Dominator 88 Café Racer 496cc,** excellent condition.
**£2,000–2,500** *LF*

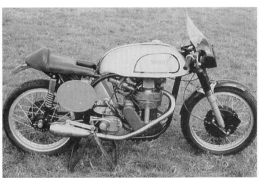

**1961 Norton Manx 40M 350cc,** frame and paintwork in good condition.
**Est. £12,000–13,000** *BKS*

**1961 Norton Navigator 350cc.**
**£900–1,400** *BLM*
*Roadholder heavyweight forks and Dominator front brakes were both features of this model.*

**1961 Norton Dominator 99 600cc,** with slim line frame.
**£2,500–3,500** *BLM*

**1961 Norton 350cc,** slim line frame and road holder forks.
**£2,000–2,500** *BLM*

1961 Norton 99SS 597cc, in excellent condition.
£2,800–3,200   *S*

**1962 Norton 40 Manx Racer 348cc,** single cylinder, DOHV, 4-stroke.
£14,000–15,000   *C(A)*

**1962 Norton Dominator 650cc,** touring version, restored to original condition.
£3,000–4,000   *PC*

**1962 Norton Jubilee 249cc,** OHV twin, restored to original condition.
£1,250–1,750   *NOC*

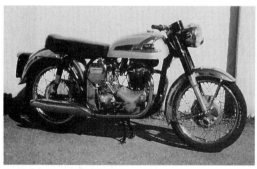

**1963 Norton 650SS 646cc,**
fully restored.
£3,000–3,200   *MR*

**1963 Norton 650SS 646cc.**
£2,800–3,200   *S*

**1970 Norton Fastback Commando 750cc.**
£2,700–3,100   *BLM*

**1969 Norton Mercury 648cc,** good condition.
£3,000–3,500   *NOC*

*r.* **c1971 Norton Commando Replica Production Racing 750cc.**
Est. £4,000–5,000   *BKS*

**1971 Norton Commando 750cc.**
**£2,700–3,000** *BLM*

*Unusual Café Racer.*

**1971 Norton Norvil Commando 745cc,**
good condition.
**£2,800–3,200** *S*

**1972 Norton Commando 828cc,**
near original condition.
**£2,500–3,000** *NOC*

**1975 Norton Commando 828cc,**
twin cylinder, original specification.
**£2,800–3,200** *BKS*

---

**Miller's is a price GUIDE
not a price LIST**

---

**1977 Norton Commando Interstate 850cc.**
**£2,500–3,000** *BLM*

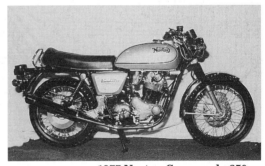

**1977 Norton Commando 850cc,**
electric start.
**£2,800–3,200** *NOC*

---

# NSU *(German)*

**1956 NSU Max OHC 247cc,** in racing trim.
**£2,200–2,500** *S*

**c1955 NSU Special Max 247cc,** OHC.
**£2,000–3,000** *AtMC*

# OEC *(British)*

**c1936 OEC-AJS 500cc,** with duplex steering frame, very rare motorcycle.
**£4,000–4,200** *BLM*

**c1939 OEC-AJS 1000cc,** V-twin, duplex steering frame.
**£4,000–5,000** *BLM*

# OK-SUPREME *(British)*

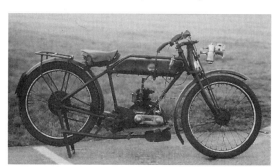

**1927 OK-Supreme 275cc,** unrestored.
**£600–800** *S*

*The manufacturers Humphries & Dawes Ltd., of Hall Green, commenced making OK motorcycles in 1899, later moving to Bramley Street and then Warwick Road, Greet, under the OK-Supreme Motors Ltd., name when Velocette took over the Hall Green factory. During the 1920s the company used engines by Blackburne and JAP with both side valve and overhead valve types, ranging from 2½ to 5hp.*

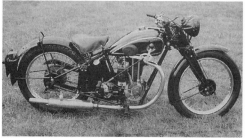

**1933 OK-Supreme 250cc,** with JAP engine, braced forks for extra strength and rigidity.
**£900–1,000** *BKS*

*This motorcycle is a descendant of the late 1920s TT winner.*

**1934 OK-Supreme Flying Cloud 245cc,** with JAP engine, professionally restored.
**£2,300–2,800** *BKS*

**1937 OK-Supreme 350cc.**
**£3,000–3,500** *PM*

**1941 OK-Supreme 350cc,** all original, in running order.
**£2,000–2,500** *S*

*Built for appraisal by the War Office, this motorcycle was the last road model OK-Supreme produced.*

# OSSA *(Spanish)*

**1974 OSSA Trials 250cc,** Renthal handlebars, Preston Petty mudguards, in running order.
**£250–350** *S*

*This trials motorcycle has been regularly used and is eligible for pre-1975 competition.*

# PANTHER (P & M)
## *(British)*

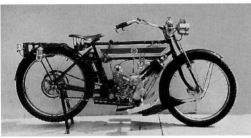

**1912 Panther P & M 500cc.**
**£8,500–9,500** *VER*

*This motorcycle is the same model as those used by the Royal Flying Corps. in WWI.*

**1937 Red Panther 248cc,** sloper single cylinder, enclosed valves, 3-speed gearbox.
**£850–1,000** *S*

**1950 Panther 350cc.**
**£1,400–1,800** *BLM*

**1956 Panther Model 65 250cc.**
**£1,000–1,100** *BLM*

**1962 Panther Model 120 645cc.**
**£1,800–2,200** *BLM*

**1934 Panther Model 100 Red Wing 600cc.**
**£3,000–4,000** *S*

*This is an ex Gangbridge machine*

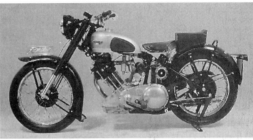

**1949 Panther M100 Sloper 598cc,** single cylinder OHV 4-stroke, bore and stroke 87 x 100mm, 23bhp.
**£3,500–4,500** *C(A)*

*Before the motor car became more universally affordable a common form of transport was the motorcycle and sidecar combination. The Sloper with its enormous torque at low revs was an ideal machine for this purpose. The machine above is a very rare single port model.*

**1951 Panther 100 600cc.**
**£2,000–3,000** *BLM*

**1964 Panther M120 Sloper 645cc,** single cylinder OHV 4-stroke, bore and stroke 88 x 106mm, 27bhp.
**£1,800–2,200** *C(A)*

*Due to stiff competition from the cheaper motor cars the M120 ceased production in 1967.*

## PARILLA *(Italian)*

**1960 Parilla Fox Racer 174cc,** single
cylinder, high pushrod OHV 4 stroke, bore and
stroke 59.8 x 62mm.
**£4,500–5,500**  *C(A)*

*Parilla produced these lightweight machines in
limited numbers for Italian style street racing.*

## PREMIER *(British)*

**1914 Premier 3½hp,** Armstrong 3-speed hub gear,
restored by Tennant-Eyles, good condition.
**£6,000–7,000**  *BKS*

## PUCH *(Austrian)*

**1956 Puch 250SGS 247cc,** split single
2-stroke, side mounted carburettor, bore and stroke
45 x 78mm, 16.5bhp, pressed steel frame.
**£1,500–2,000**  *C(A)*

## RADCO *(British)*

**1929 Radco Lightweight,** Villiers 147cc engine,
3-speed, finished in silver and black.
**£900–1,200**  *S*

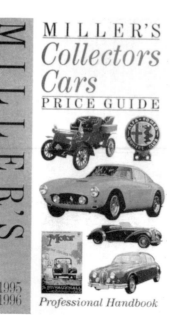

# RALEIGH *(British)*

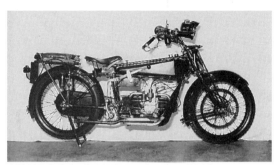

**1921 Raleigh,** flat twin engine, leg shields, chain drive, all original.
**£7,500–8,000** *HRM*

*Only six of these machines are known to exist.*

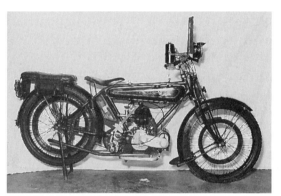

**1923 Raleigh Model 2 2¾hp,** 3-speed belt/chain drive, belt rim rear brake, dummy rim front brake, new inner tube and replaced belt otherwise original.
**£3,000–4,000** *HRM*

**1924 Raleigh 3hp,** belt and chain drive, rear belt rim brakes, completely restored.
**£4,000–5,000** *HRM*

**1923 Raleigh 2¾hp 350cc,** flat tank.
**£3,000–3,500** *BLM*

**1923 Raleigh Model 5 2¾hp 350cc.**
**£2,700–3,000** *BMM*

**1926 Raleigh Model 5 2¾hp 350cc.**
**£2,200–2,400** *BMM*

**1927 Raleigh 2¾hp 350cc.**
**£2,800–3,400** *BLM*

*l.* **1928 Raleigh 495cc,** single cylinder side valve 4-stroke, bore and stroke 79 x 101mm.
**£3,500–4,000** *C(A)*

*A by-product of the Raleigh motorcycle factory was the sale of gearboxes under the Sturmey-Archer name.*

# RAVAT *(French)*

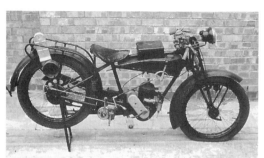

**c1924 Ravat Type P4 Solo 173cc,** complete, in original condition.
**Est. £800–1,000**  *BKS*

# REX-ACME *(British)*

**1926 Rex-Acme Model K6 350cc,** side valve Bluckburne unit, Burman 3-speed gearbox, restored, very good condition throughout.
**Est. £1,400–1,800**  *BKS*

# RICKMAN *(British)*

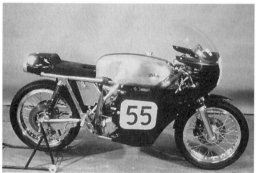

**1966 Rickman G50 Matchless Metisse 496cc,** engine re-built by Bob Newby, Marzocchi rear suspension units, 5-speed Quaife gearbox re-built by Mick Hemmings, new rims and tyres to wheels, good condition, race prepared.
**Est. £12,000–14,000**  *S*

**1969 Rickman Metisse 649cc,** Triumph 649cc twin cylinder pre-unit engine, twin carburettors, nickel-plated frame, restored, excellent condition.
**£2,300–2,800**  *BKS*

**1973 Rickman Triumph TR6 650cc.**
**£3,000–4,000**  *BLM*

**1969 Rickman Triumph Metisse 649cc,** Rickman chassis, 649cc pre-unit T120 Bonneville engine, restored.
**£3,000–3,500**  *BCB*

**c1970 Rickman Triumph Tiger 100 490cc.**
**£1,700–2,000**  *BLM*

**1974 Rickman Metisse Bonneville Twin Cylinder Unit-Construction Model 649cc,** standard factory service replacement engine and gearbox.
**£3,000–3,300**  *PS*

# ROYAL ENFIELD *(British 1901–70, Indian 1955–)*

The slogan most often quoted for the Royal Enfield marque is 'made like a gun', its company logo incorporating a small field gun as a trademark.

Royal Enfield experimented with motorcycle engines in 1901 but it was not until 1911 that a complete machine was offered for sale, this being a 425cc MAG V-twin. From 1912–20 a variety of motorcycles were produced but it did not manufacture their own engines until 1920. By the late 1920s only Royal Enfield engines were used in all their models.

Prior to WWII Royal Enfield specialised in motorcycles that were built mainly for export, such as the luxurious 1140cc side valve V-twin. After the war it built the Prince, a small two-stroke single, the 248cc Crusader, the 346 and 499cc Bullet and a number of large capacity overhead valve twins including the Meteor, Constellation and later the Interceptor.

In 1963 the company was sold to the E. & H. P. Smith engineering group but this did little to halt the decline of the famous Redditch, Worcestershire, marque. After a series of cut backs and financial problems the company was once again to change hands, this time to NVT at Bradford-on-Avon. Enfield Interceptor engines were also used in the Rickman Enfield of the early 1970s.

Meanwhile a branch of Royal Enfield had been established in Madras, India, in the 1950s, to concentrate on production of the Prince 2-stroke and later the Bullet single. This branch still exists today and Indian-built Enfields have been imported to Britain since the late 1970s.

**1921 Royal Enfield Model 200 2½hp 225cc,** 2-stroke power unit, lightweight, fully restored in 1993.
**£1,400–2,000** *BKS*

**1922 Royal Enfield 225cc,** 2-stroke, very good condition throughout.
**£1,200–1,800** *C*

**c1930 Royal Enfield 250cc.**
**£1,500–1,800** *BLM*

**1939 Royal Enfield Model G 350cc.**
**£1,800–2,000** *BLM*

**1943 Royal Enfield WD/CO.**
**£1,000–1,200** *MVT*

**1946 Royal Enfield WD/CO 350cc.**
**£1,900–2,000** *BLM*

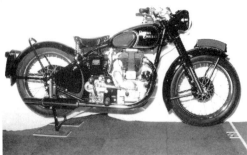

**1951 Royal Enfield 350cc,**
restored, good condition.
**£2,000–2,500** *PC*

**1948 Royal Enfield Model G 346cc,**
re-built to original specification.
**£1,800–2,200** *S*

**1952 Royal Enfield J2 499cc,** enclosed valve
gear, finished in black, good condition.
**£1,500–2,000** *S*

**1952 Royal Enfield J2 499cc,** twin port head,
finished in black, excellent condition throughout.
**£1,800–2,200** *S*

**1953 Royal Enfield Bullet Trials 346cc.**
**£1,600–1,800** *S*

**1955 Royal Enfield Bullet Trials 499cc.**
**£1,200–1,500** *MR*

**1959 Royal Enfield Meteor Minor**
**500cc,** complete and original.
**Est. £1,600–1,800** *BKS*

**1959 Royal Enfield Crusader, 246cc,** complete
and mostly original, requires restoration.
**£220–320** *S*

**1959 Royal Enfield Bighead Bullet 499cc,**
OHV single.
**£3,000–3,500** *AtMC*

**c1960 Royal Enfield Meteor Minor 496cc,**
OHV parallel twin.
**£2,000–2,500** *AtMC*

**1960 Royal Enfield Meteor Minor 496cc.**
**£2,000–2,500** *BLM*

**1961 Royal Enfield Crusader 250cc,**
requires restoration.
**£300–400** *CVPG*

**1961 Royal Enfield Meteor Minor 500cc,**
vertical overhead valve, 4-speed posi-stop gearbox,
fully restored.
**£2,300–2,800** *BKS*

**1961 Royal Enfield Crusader Sports 250cc.**
**£1,200–1,400** *BLM*

**1962 Royal Enfield Super Five 250cc,** restored
to original specification, very good condition.
**£2,000–2,500** *BKS*

**1962 Royal Enfield Crusader Trials 250cc,**
new gearbox, sleeve gear and electrics.
**£900–1,000** *BKS*

*r.* **1962 Royal Enfield**
**Continental GT 250cc.**
**£1,500–2,000** *BLM*

**1964 Royal Enfield Turbo Twin 249cc,**
Villiers 4T engine.
£800–1,100  *BLM*

**1965 Royal Enfield Continental GT 248cc.**
Est. £1,200–1,400  *S*

**1965 Royal Enfield Continental GT,**
good condition.
Est. £1,600–1,750  *S*

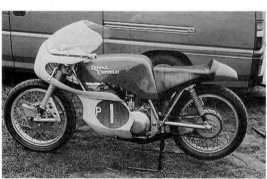

**1965 Royal Enfield GP5 Racer 246cc,**
single cylinder 2-stroke with 5-part
cylinder, 34bhp at 8000rpm.
£2,400–3,100  *PC*

# ROVER *(British)*

**1902/3 Rover 2¼hp Solo,** single speed, belt driven
with pedal assistance, restored.
£10,000–10,500  *AtMC*

*Motorcycle production at The Rover Cycle Co. Ltd.
works at Coventry commenced in 1902, alongside
the already successful bicycle business and
continued until 1925.*

**1967 Royal Enfield Continental GT,**
excellent condition.
Est. £1,200–1,400  *S*

# ROYAL-RUBY (British)

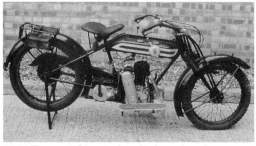

**1920 Rover 4½hp,** single cylinder side valve
engine, rear drum brake, spring link drive as fitted
to TT models, Ariel gearbox with silent tooth drive,
black and silver livery, good condition throughout.
£3,000–3,500  *BKS*

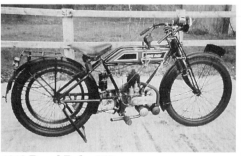

**1913 Royal Ruby 275cc.**
£3,000–4,000  *BLM*
*A rare London to Brighton runner.*

# RUDGE *(British 1911–1940)*

Although famous for bicycles, Dan Rudge did not begin motorcycle manufacture until the launch of the single cylinder 500cc IOE engined model in 1911. That same year, Rudge's works rider Vic Surridge broke several records at Brooklands, including the 'hour' at over 65mph.

The Company contested the major races of the day and also entered a team of riders in the 1911 TT, but after Surridge was killed in a practice, the team withdrew. Rudge returned to the Isle of Man in 1913 when Ray Abbott finished second, while in 1914 Cyril Pullin had the distinction of winning the only two-day Senior TT ever staged.

Their best-selling machine of the veteran and vintage period was probably the 499cc Multi (for multi-gear, not cylinders) which was manufactured both before and after WWI. In 1924 their new 4 valve 349 and 499cc singles, with chain drive and 4 speeds, proved to be a sales success.

By the late 1920s, Rudge had made a return to racing, which was instigated by managing director J. V. Pugh with George Hack as designer and development engineer. This partnership soon produced some outstanding results, witnessed by wins in the 1928 and 1929 Ulster GPs by Graham Walker (father of Murray). Besides Walter Hanley's famous 1930 Senior TT victory, Rudge were also a leading force in the Lightweight (250cc) class. As an interesting aside, Enzo Ferrari ran Rudge machines under his own Scuderia Ferrari banner in the early 1930s, before his interest centred on the Alfa Romeo and then Ferrari cars. After changing hands in 1936 in a buy-out involving the electronics giant EMI, the factory was moved from Coventry to Hayes, Middlesex. The Rudge Ulster is probably the marque's most famous model, but even this was not enough to save this once great name from extinction.

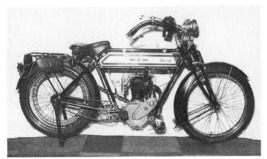

**1911 Rudge, 499cc.**
**£5,000–5,500** *VMCC*

**1911 Rudge 500cc.**
**£6,000–6,500** *VER*

*This motocycle was the fourth off the production line.*

**1930s Rudge Special 500cc.**
**£3,000–3,500** *BLM*

**1929 Rudge Whitworth 340cc.**
**£3,000–3,500** *PC*

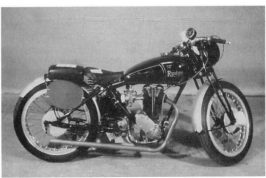

**1930s Rudge Special 500cc.**
**£2,000–3,000** *BLM*

**1934 Rudge Solo Racing 250cc.**
**Est. £2,000–3,000** *S*

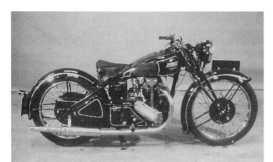

**1936 Rudge Ulster 500cc,** OHC,
twin port single, 4 valve.
**£5,500–6,500**  *S*

**1937 Rudge Rapid 245cc,** OHC, restored to
original condition.
**£0,000–7,500**  *REC*

*This motorcycle has won many concours awards.*

**1937 Rudge Rapid 249cc,** girder forks, rigid
frame at the rear, pillion seat, in running order.
**£1,400–1,600**  *BKS*

**1937 Rudge Rapid 250cc.**
**£2,000–2,500**  *BLM*

**1937 Rudge Special 499cc,** 4 valve head, full
enclosed valve gear, partly restored.
**£4,500–4,750**  *VER*

**1937 Rudge Rapid 250cc,** complete, correct and
original, running well, Swansea V5.
**Est. £2,000–2,500**  *S*

**1938 Rudge Special,** 4 valve radial head
configuration, enclosed valve gear, valanced
mudguards, to original specification, in good
condition, buff log book and Swansea V5.
**£3,800–4,200**  *S*

# SCOTT *(British)*

**1920 Scott,** in original condition, some repainting, V5 document and original registration number.
**Est. £3,750–4,250** *S*

**1925 Scott TT Racing 600cc Replica,** water-cooled siamesed twin inclined cylinder engine, 2-speed gearbox, large front brake and low slung frame, good condition, engine in good running order.
**Est. £4,000–4,500** *S*

*This machine is very rare.*

**1925 Scott 486cc,** siamesed 2-stroke twin, 2-speed, water-cooled, painted in black and ruby livery, Swansea V5.
**Est. £3,500–4,000** *BKS*

**1927 Scott Flying Squirrel 492cc,** 2-speed, finished in Scott purple livery, engine reconditioned 25 years ago.
**£3,000–4,000** *BKS*

**1929 Scott Flying Squirrel 496cc,** water-cooled parallel-twin 2-stroke power unit.
**Est. £4,000–5,000** *BKS*

**1929 Scott Squirrel 299cc,** single cylinder.
**£2,000–2,500** *BLM*

**1929 Scott Flying Squirrel.**
**£4,000–4,600** *VER*

**1930 Scott 596cc,** water-cooled twin 2-stroke siamesed inclined twin cylinder engine, girder forks and rigid frame at rear.
**£1,300–1,800** *S*

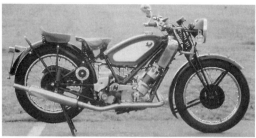

**1938 Scott 496cc,** siamesed water-cooled 2-stroke engine, mechanically in excellent condition, good painted black finish, Swansea V5.
**£3,500–4,000** *S*

**1937 Scott Squirrel 496cc,** twin-cylinder water cooled engine, restored.
**Est. £3,400–3,600** *BKS*

*l.* **1930 Scott Power Plus 596cc,** re-built to factory specification, original registration number and original log book.
**£3,500–4,000** *BKS*

*The Scott factory enjoyed an early reputation for success in racing certainly due to the engineering genius of Alfred Angus Scott who died of pleurisy in 1923. He pioneered the success of 2-stroke racing and introduced the rotary valve and water-cooling for racing engines, ideas decades ahead of their time and re-used today. Their finest time was before WWI, but the factory never quite regained its pre-war dominance.*

# SEELEY *(British)*

**1947 Scott 596cc,** 2-stroke twin, rigid frame allied to Dowty hydropneumatic forks, finished in black and maroon, good overall condition.
**Est. £3,000–4,000** *BKS*

**1972 Seeley G50 Mk 4 Racer 496cc.**
**£12,000–15,000** *S*

**1978 Seeley 750cc,** Honda 750cc engine bored to 810cc, full fairing, alloy tank, frame and sports seat, forks with double discs, Lester alloy wheels, painted white.
**Est. £3,000–3,250** *S*

# SILK *(British)*

**1970s Silk-Scott 700S,** 700cc 2-stroke twin engine, built by Derby-based Silk Engineering.
**£3,000–4,000** *BLM*

# SUNBEAM *(British 1912–1957)*

Sunbeam was famous for both its pedal cycles and motorcycles. The original owners were John Marston of Wolverhampton, but after WWI, the company was sold to Nobel Industries (which later became the corporate giant ICI). It was under their management that Sunbeam became known as a quality product in terms of performance and finish.

J. E. Greenwood was Chief designer for Sunbeam from 1912 to 1936. He was highly respected within the industry and when ICI sold the Sunbeam name to the AMC Group in 1936 Greenwood did not transfer to AMC, which marked the end of Sunbeam's racing involvement.

The new owners moved production to London where new 250, 350 and 500cc models, with Collier designed high camshaft engines, were shown at the 1938 London Show. These were to be short lived as the war intervened and Sunbeam was once again sold, this time to BSA.

Post-war, Ehrling Poppe designed the S7. This was a luxury 497cc all-alloy OHC twin, with in-line mounting, four-speed gearbox with shaft final drive. Although a comfortable and pleasant tourer, the S7 did not prove very popular, and the S8 sports version also failed to meet sales targets. BSA finally axed Sunbeam's motorcycle production in 1957, although the name continued on bicycles and, for a short time, on the BSA Sunbeam scooter with basically a 175cc Bantam engine.

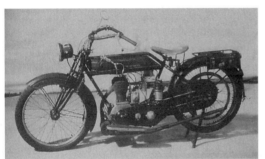

**1916 Sunbeam 500cc,** single cylinder side valve, 3-speed countershaft gearbox, chain final drive running in an oil bath, fully restored, in excellent condition, Swansea V5 and old buff log book.
**£4,500–5,500** *S*

**c1918 Sunbeam TT Replica 500cc.**
**£3,000–3,500** *BLM*

**1929 Sunbeam Model 5 493cc.**
**£2,200–2,800** *BLM*

**1919 Sunbeam 3½hp,** complete, unrestored and with documentation.
**£3,500–4,500** *BKS*

**1929 Sunbeam Model 80 346cc,** single cylinder OHV engine, 3-speed gearbox, fully re-built condition, in traditional Sunbeam black and gold livery.
**£5,000–6,000** *BKS*

**1933 Sunbeam Model 14 247cc,** longstroke 59 x 90mm, modified model.
**£2,500–3,500** *BKS*

*r.* **1937 Sunbeam Model 9 Mk VI 493cc,** original condition, history file, Swansea V5 and old style log book.
**£3,500–4,000** *S*

**1939 Sunbeam B24 348cc,** high camshaft model.
£1,200–1,600 *AT*

**1939 Sunbeam B24 348cc,** Swansea
V5 and current MOT.
**Est. £3,000–3,400** *S*

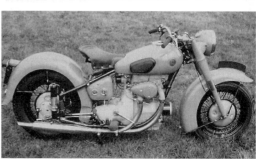

**1950 Sunbeam S7 487cc,** shaft drive, chain
driven, OHC, two-into-one exhaust, fluted silencers,
deeply valenced mudguards, traditional light green
paintwork, good condition throughout.
**Est. £2,500–3,000** *BKS*

**1940 Sunbeam C23 Highcam 245cc,** painted
in Sunbeam livery of gold and black, Swansea
V5 and old buff log book.
**£2,000–2,500** *S*

*l.* **1951 Sunbeam
S8 487cc,**
**£2,500–3,500**
*CMAN*

**1957 Sunbeam S8 487cc,** in-line twin, OHC
engine, the Sports version of S7 Tourer, painted
silver grey, restored.
**£2,000–3,000** *BKS*

*l.* **1955 Sunbeam S8 497cc.**
**£1,200–1,600** *AT*

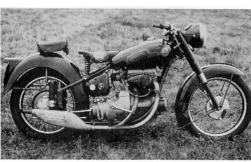

**1957 Sunbeam S8 487cc,** valenced mudguards,
painted black, good condition.
**£1,500–2,000** *BKS*

**1959 Sunbeam S8 487cc,** twin shaft drive on
OHC, carefully restored, finished in silver grey,
concours condition, excellent running order.
**£4,200–4,500** *S*

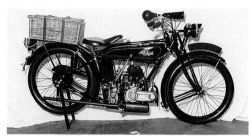

**1924 Raleigh Model 14 250cc,** 3 speed gearbox, 2 rear brakes, no front brakes, all chain drive, 24in wheels, fully restored.
**£2,500–3,000** *HRM*

**1926 Raleigh Model 16 348cc,** all chain drive, 26in wheels.
**£3,000–3,500** *HRM*

**1931 Raleigh 298cc.**
**£1,300–1,800** *S*

**1921 Royal Enfield 2¼hp Lightweight 200cc.**
**Est. £2,000–2,500** *BKS*

*The first 2¼hp 2-stroke single was unveiled in 1914. WWI curtailed its civilian sales but in 1920 one was entered for the ACU Six Days' Trial, winning a gold medal.*

**1956 Royal Enfield Clipper 246cc,** OHV single, re-built.
**£500–700** *BCB*

**1938 Royal Enfield Model J 499cc,** coil ignition, 4-speed posi-stop gearchange, concours condition.
**£2,500–3,000** *S*

**1957 Royal Enfield Meteor 700 OHV Vertical Twin 692cc,** restored.
**£1,500–2,000** *PS*

**c1953 Royal Enfield 500 Twin 497cc,** OHV parallel twin.
**£3,500–4,000** *AtMC*

*Royal Enfield's 500 Twin first appeared in 1949.*

*l.* **1965 Royal Enfield GT Continental 249cc,** OHV single, concours condition, restored.
**£2,000–2,500** *BCB*

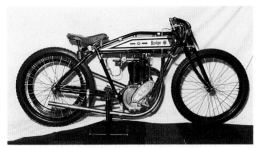

**1913 Rudge Brooklands Racing 499cc.**
**£7,500–8,500** *VMCC*

*This motorcycle is the only remaining example of this model, which was first to achieve 80mph.*

**1929 Rudge Whitworth 250cc,** black with gold lining, hand change gearbox, single sprung saddle and sport type mudguards.
**£2,000–2,500** *S*

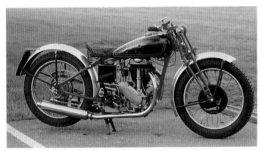

**1937 Rudge Ulster,** polished alloy mudguards, black and red paintwork, good cosmetic condition.
**£1,500–2,000** *S*

**1927 Scott Flying Squirrel 596cc,** restored.
**£3,000–3,500** *BCB*

**1925 Rudge 350cc,** all original, unrestored, all mechanical elements, frame, paintwork and tyres in good condition.
**Est. £3,000–3,500** *BKS*

**1936 Rudge Ulster GP 500cc,** restored to concours and original condition.
**£10,000–11,000** *REC*

*This motorcycle won the post vintage category at the 1995 Stafford Classic Motorcycle Show.*

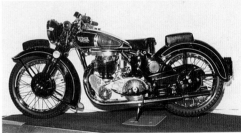

**1939 Rudge Special 500cc,**
4-valve twin port single cylinder engine.
**£8,000–8,500** *REC*

**1937 Scott Flying Squirrel 498cc,** water-cooled siamesed 2 stroke twin engine, complete and original, unrestored condition.
**£2,600–3,200** *S*

*l.* **1932 Scott Solo 2 Stroke Twin 498cc,** brass water tank and petrol/oil tank configuration, TT modifications, good condition.
**£2,200–3,000** *S*

**c1923 Sunbeam Sport.**
**£6,000–7,000** *AtMC*

**1932 Sunbeam 493cc,** restored in 1991 after
being laid up from 1957.
**£2,900–3,500** *PC*

**1938 Sunbeam Model 8 346cc,**
OHV single cylinder engine,
restored to excellent condition.
**Est. £3,500–3,750** *S*

**1933 Sunbeam Model 9 493cc OHV,**
saddle tank and fishtail exhaust, extensively
rebuilt and in excellent condition.
**£2,800–3,200** *S*

**1950 Sunbeam Model S7 487cc,**
balloon tyres and shaft drive.
**£3,000–3,500** *C(A)*

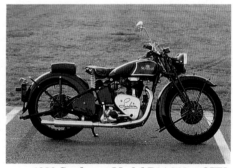

**1939 Sunbeam Model B24 348cc OHV,**
wheels rebuilt, renewed cables, rubbers
and fork spindles.
**£2,600–3,200** *S*

**1951 Sunbeam Model 57 Deluxe 497cc,**
OHC, in-line twin, shaft drive, fully restored.
**£3,400–3,600** *SOF*

**1955 Sunbeam Model S7 Twin 500cc,**
chain drive, overhead camshaft, with shaft
drive to the rear wheel, in excellent condition.
**£4,200–4,800** *S*

**1921 Sparkbrook Model A 269cc,** single cylinder, 2-stroke Villiers engine, restored to excellent condition.
**£2,000–2,500** *PS*

**1968 Suzuki T20 Sports 246cc,** 2-stroke twin cylinder engine, racing linings installed in the front brake, upswept exhaust system.
**Est. £2,000–2,200** *S*

**1972 Suzuki TR500 Ex-Works Racing 492cc,** water-cooled engine, hydrolically operated disc brakes front and rear, in ready-to-race condition.
**£8,800–9,200** *S*

**1974 Suzuki 500 Ex-Works Racing 500cc,** twin cylinder engine, full racing fairings, excellent condition.
**Est. £9,000–11,000** *BKS*
*This machine was raced by Barry Sheene.*

**1976 Suzuki TR750 Ex-Works Racing 750cc,** close ratio speed works gearbox, magnesium carburettors and clutch, racing fairing, re-built.
**Est. £11,000–13,000** *BKS*

**1975 Suzuki TR750 Ex-Works Championship Racing 750cc,** restored, in excellent condition.
**Est. £17,000–19,000** *S*
*Raced by Barry Sheene in 1975 at the Race of the Year at Mallory Park.*

**1977 Suzuki GT550 550cc,** 2-stroke, 3-cylinder, restored and fitted with twin front disc brakes.
**£1,400–2,000** *PS*

**1979 Suzuki RG500 MkIV Racing,** full racing trim, with fairings and windshield.
**Est. £4,600–5,000** *S*
*This motorcycle was originally purchased by ex-'Beatle' George Harrison.*

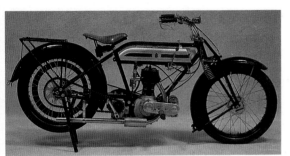

**1924 Triumph Model SD 549cc,** single cylinder, side valve, 4-stroke, 4hp.
**£3,000–4,000** *C(A)*
*Called 'The Trusty Triumph' by WWI despatch riders this motorcycle was built in many different guises.*

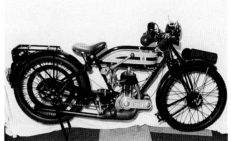

**1926 Triumph Model P 500cc.**
**£3,000–3,500** *PVE*
*This model was at the cheap end of the Triumph range.*

**1913 Triumph 499cc,** single cylinder, side valve, air-cooled, 4-stroke engine, 3-speed hub gearbox, belt drive and pedal assistance.
**Est. £4,900–5,500** *BKS*

**1935 Triumph Model L2/1 249cc OHV.**
**£900–1,500** *PS*

**1937 Triumph 5T Speed Twin 499cc,**
V-twin, OHV, 4-stroke, 27bhp.
**£5,500–6,500** *C(A)*

**1938 Triumph Tiger 80 343cc,** high level exhaust system, with tank top instrument panel, in good condition.
**£1,300–2,000** *S*

**1947 Triumph 5T Speed Twin 499cc,** restored, in good condition throughout.
**Est. £3,500–4,000** *S*

**c1949 Triumph TR5 Trophy Trials 499cc,** V-twin OHV sand cast cylinder head and barrel engine, re-built.
**Est. £2,500–3,500** *PS*

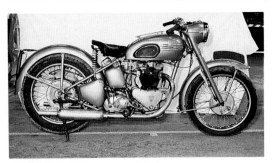

**1950 Triumph 6T Thunderbird 649cc,**
frame 9176, engine no. 9176N.
**£2,500–3,500**  *PS*

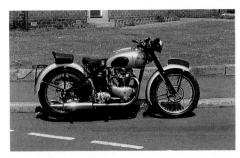

**1951 Triumph Tiger 100,** sprung rear hub,
telescopic front forks.
**£2,800–3,500**  *PM*

**1953 Triumph 5T Speed Twin 499cc,**
in good original condition.
**£1,700–2,500**  *MR*

**1955 Triumph Tiger 100 499cc,** OHV twin,
swinging arm frame, restored to concours condition.
**£4,000–5,000**  *BCB*

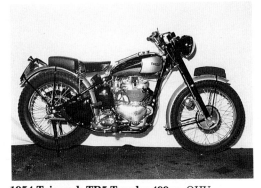

**1954 Triumph TR5 Trophy 499cc,** OHV.
**£4,500–5,500**  *ST*

*The original TR5 Trophy was built with the
International Six Days Trial very much in mind.*

**1955 Triumph Thunderbird 649cc,** swinging
arm rear suspension, coil ignition and alternator,
in partially restored condition.
**Est. £3,000–4,000**  *S*

**1961 Triumph 6T Thunderbird 649cc,**
a 'Bath-tub' model.
**£3,750–4,100**  *AT*

**c1961 Triumph Trophy TR6 649cc,** pre-unit OHV
parallel twin.
**£5,000–6,000**  *AtMC*

**1969 Triumph T120 Bonneville 649cc.**
£3,000–4,500  *ALC*

**1964 Rickman Triumph Mettisse Scrambler 490cc.**
£4,000–4,500  *AtMC*

**1964 Triumph Model 21 350cc,**
original metalwork.
**Est. £1,000–1,200**  *S*

**1969 Triumph T150 Production Racer 740cc,**
tuned motor, Fontana front brake,
original condition.
£4,500–5,000  *ST*

*Supplied for TT marshalls 1969–74, being only
one of 4 made.*

**1971 Triumph SS Blazer 250cc,**
excellent fully restored condition.
£800–1,200  *MR*

**1975 Triumph Trident T150 740cc,**
US specifications, excellent condition,
fully restored throughout.
£4,500–5,000  *PC*

*This motorcycle is one of 11 made.*

**1979 Triumph T140E Bonneville 747cc,**
OHV, vertical twin engine, non-standard instruments
and fittings, external oil filter, twin front disc brakes,
some wear in valve guides.
£2,800–3,000  *PS*

**1980 Triumph T140E,** fitted with
Sigma panniers, top box fairing, limited
edition, good mechanical condition.
**Est. 2,200–2,600**  *S*

**1925 Velocette Ladies Model EC 250cc,**
2-stroke engine, unrestored, very rare original
**£2,000–3,000** *Vel*

**1931 Velocette Model GTP 248cc,**
single cylinder, 2-stroke, twin port.
**£3,500–4,000** *C(A)*

**1936 Velocette Model KTS 348cc,** single cylinder,
OHC, 4-stroke, 29bhp.
**£4,000–4,500** *C(A)*
*This is the touring version of the KSS racer with valenced
guards, produced until 1939*

**1935 Velocette KTT Racer 349cc.**
**£4,000–5,000** *PM*

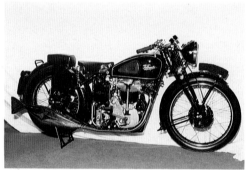

**1937 Velocette KSS MkII 349cc,** OHC, restored.
**£5,000–6,000** *Vel*

**c1945 Velocette Model MDD 350cc,**
WD model, OHV engine, rigid rear
frame and girder forks.
**£900–1,400** *S*

**1948 Velocette KTT MkVIII 348cc,** OHC engine,
Girling type rear spring dampers, excellent
mechanical condition.
**£11,000–12,000** *BKS*

**1952 Velocette LE200 192cc,**
horizontally opposed twin, side valve,
hand gear change and hand starter.
**£650–700** *PS*

**1956 Velocette Viper 349cc.**
£3,000–3,500 *Vel*

*This was the 219th Viper built and was later converted to Thruxton specifications.*

**1958 Velocette Viper 349cc,** with clubman pattern front brake, polished front mudguard, good condition throughout.
£2,700–3,200 *S*

**1960 Velocette Viper Clubman 349cc,** single cylinder OHV 4-stroke engine, 28bhp.
£6,500–7,000 *C(A)*

**1961 Velocette Viper 349cc,** V-line, in excellent mechanical condition, new tyres and battery, chromium plated mudguards and engine cowlings, all black/chrome in good condition.
£2,500–3,000 *MR*

**1965 Velocette LE 192cc.**
£800–1,000 *PM*

**1966 Velocette Thruxton 499cc,** GP carburettor removed, finished in black, good condition.
Est. £5,000–6,000 *S*

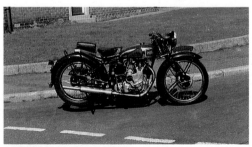

**1938 Vincent HRD 499cc.**
£6,000–6,500 *PM*

**1952 Vincent Rapide 998cc,** some replacement parts, converted 12 volt coil ignition.
£7,800–8,200 *MR*

**1964 Yamaha Model YDS 3,** restored and in good condition.
**Est. £2,000–2,500** *S*

**1969 Yamaha TR2 348cc,** twin cylinder, 2-stroke engine, restored to good original condition.
**£3,800–4,200** *S*

**1972 Yahama TR3 Racing 349cc,** air-cooled engine, 6-speed gearbox, square barrels, 58bhp at 9000rpm, left hand tank cap, restored to original specification.
**£3,200–3,800** *S*

**1971 Yamaha TR2B Racer 348cc,** twin cylinder 2-stroke, 56bhp at 10,000rpm, cap on left hand side of tank support rod for carburettors.
**£4,000–4,500** *C(A)*

**1973 Yamaha TZ350A Racing 347cc,** 60bhp at 9500rpm.
**£4,000–4,500** *S*

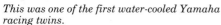

*This was one of the first water-cooled Yamaha racing twins.*

**1976 Yamaha TZ Vesco 750 Racing 750cc,** 4-cylinder engine, close ratio racing gearbox, full racing fairing, good condition throughout.
**Est. £8,000–9,000** *BKS*

**1980 Yamaha TY175 Trials.**
**£1,100–1,600** *S*

**1916 Zenith Gradua,** JAP V-twin engine, apart from exhaust pipes and thrust ball race on the engine pulley, to original specification, in good condition.
**£5,000–5,500** *S*

**1957/71 Norton Triumph (Triton) 649cc Café Racer,** 1971 TR6R 650cc engine, 1957 Norton wide line frame with a Manx swinging arm. **Est. £2,000–2,200** *S*

**1979 Triumph T140 Bonneville Special OHV Twin 744cc,** rebuilt.
**£2,000–2,500** *BCB*
*This motorcycle has an export small tank, and non-standard short Megatron silencers. All specials had alloy wheels.*

**1960 Triton pre-unit 649cc,** Triumph engine, Norton Dominator cycle parts.
**£1,800–2,300** *PS*

**1940/60s Pegasus Vincent 1459cc Drag Racing and Sprint Special,** gearbox casing used as an oil tank.
**£10,000–15,000** *BKS*

**c1970 Indian Velocette 499cc.**
**£7,000–7,500** *AtMC*

*Floyd Clymer was behind the Indian Velo 500. It used the 499cc Thruxton engine and Italian cycle parts, with an American name.*

**c1963 Triton Solo Racing 649cc.**
**£1,800–2,000** *S*

**1973 Seeley Suzuki TR500 Daytona 492cc.**
**£7,000–7,500** *PC*

*The monocoque frame technology derived from aviation industry and motor racing. It was test ridden by Barry Sheene and Stan Woods, and developed by Colin Seeley.*

**1977 Seeley-Honda 750cc,** 4-cylinder Honda air-cooled engine, twin headlamps, full fairing, alloy tank, single seat, Honda front forks, 'Comstar' wheels.
**Est. £2,250–2,500** *S*

**1916 BSA 4¼hp 575cc Combination,** with carbide lighting set, finished in traditional BSA green livery, restored to good condition throughout.
**£4,500–5,000**  *S*

**1951 AJS 18S 497cc, with Tilbrook Sidecar,** single cylinder, OHV, 4-stroke engine, 21.1bhp.
**£7,000–8,000**  *C(A)*

*The Model 18S 497cc was the workhorse of AJS post-WWII range and when coupled with a sidecar could provide cheap transport for a small family.*

**1957 BMW R60 599cc with Steib Sidecar,** twin horizontally opposed OHV 4-stroke, bore and stroke 73.5 x 70.6mm, 40bhp.
**£7,000–8,000**  *C(A)*

**1925 AJS 8hp Combination,** restored to good condition.  **Est. £4,000–5,000**  *S*

*The AJS factory in Wolverhampton produced a series of motorcycle sidecar combinations from 1914.*

**1929 Indian 101 Scout 596cc Combination,** V-twin side valve 4-stroke.
**£7,000–8,000**  *C(A)*

**1949 Norton 18 490cc Combination.**
**£2,800–3,500**  *PM*

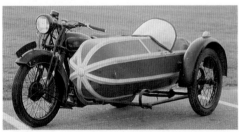

**1946 Norton Combination,** in traditional Norton livery of black with silver tank, the single seat sporting type sidecar with windshield.
**£2,100–2,600**  *S*

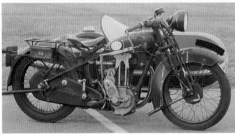

**1930 Sunbeam Model 9 Sports Combination,** hand change gearbox, girder fork and coil spring front suspension, Smith's speedometer, Lucas lighting, twin panniers, torpedo lightweight sports sidecar.
**£3,700–4,200**  *S*

**1923 Sunbeam 3½hp 500cc Combination,** single cylinder, side valve engine, acetylene lighting and period bulb horn, finished in traditional black livery, the chair with apron, screen, and red upholstery.
**Est. £6,000–8,000**  *BKS*

**1948 Corgi Scooter 98cc,** developed from the wartime Welbike, but with horizontal Excelsior Spryt engine, tubular frame, 12½in wheels, re-built.
**£350–400** *S*

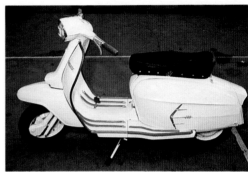

**1968 Lambretta SX200 198cc,** restored to standard original specification.
**£1,500–2,000** *PS*

**c1953 Lambretta Model F 123cc.**
**£850–1,000** *EP*

*This is a very rare model.*

**1960 Laverda Mini Scooter 49cc,** 4-stroke engine, 2-speed gearbox, 2bhp.
**£500–600** *IMO*

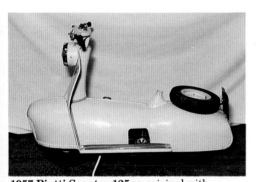

**1957 Piatti Scooter 125cc,** original with no modifications.
**£750–800** *PS*

*Imported from Italy.*

**c1948 Lambretta Model B 123cc.**
**£1,000–1,500** *EP*

**1956 Zündapp 201 Bella 199cc,** complete, restoration project.
**£800–850** *PS*

*This scooter has been imported from Germany.*

**1957 Zündapp Bella 154,**
Est. **£400–500** *BKS*

*Following the success of Lambretta and Vespa, Zündapp manufactured fashionable motor scooters with their all-enveloping coachwork.*

**1970 Ariel 3 Moped,** complete, in good condition.
**£50–100** *S*

**c1955 Heinkel Perle 49cc Moped,**
original green paintwork.
**£250–350** *S*

**1963 Honda C50 Cub Moped,**
complete unrestored condition.
**£50–100** *S*

**1962 Kerry Capitano Moped,**
telescopic front forks, swinging
arm rear end, drum brakes,
complete, good condition.
**Est. £160–200** *S*

**1963 NSU Quickly 23 49cc.**
**£400–450** *PS*

**1962 Puch Cheetah Moped,** fan-cooled
single cylinder 2-stroke engine, Earles type
front forks, pivoted fork rear suspension, dual
seat, full legshields, complete, unrestored.
**£125–250** *S*

**1959 Raleigh 49cc Runabout Moped,**
engine by Sturmey-Archer, complete
and original condition.
**£60–120** *S*

**c1980 Suzuki FZ 50 49cc Moped,**
complete with lighting set, dual seat and
luggage carrier, original specification.
**£200–300** *S*

A *Triumph Motorcycles, 1929*
**Brochure,** 10 x 7in
(25 x 17.5cm).
**£40–45** *DM*

**A copy of** *The Motor Cycle,*
for 5th December, 1929, 11 x 8in
(28 x 20cm).
**£15–25** *DM*

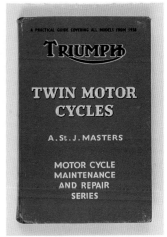

**A copy of Triumph** *Twin
Motor Cycles,* Motor Cycle
Maintenance and Repair Series,
by A. St. J. Masters, 1938–57,
7½ x 5in (19 x 12.5cm).
**£10–25** *DM*

**The Iron Redskin Indian
Motorcycle Poster,** by
Harry V. Sucher, 1977,
10 x 7in (25 x 17.5cm).
**£25–35** *DM*

**A silver cigarette case,** won by
TT rider W.L. Handley in 1925.
**£220–280** *BCA*

**A nickel plated brass
headlamp,** by P & H
Birmingham, poor
condition, 5½in
(14cm) diam.
**£23–27** *ATF*

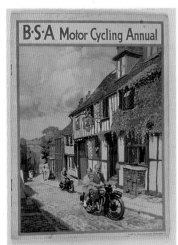

**A** *BSA Motor Cycling Annual,*
1935, 11 x 8in (28 x 20cm).
**£75–100** *DM*

**c1930 Hammond Petrol Pump.**
**£4,000–4,500** *HOLL*

**c1946 Regent Oil
Pump,** , 80in
(203cm) high.
**£400–500** *MSMP*

A collection of assorted motorcycle
tax discs, 1955, 1956 and 1959.
£8–10 each  *ATF*

A spare parts kit for motorcycle
chains, made by the Coventry Chain
Co. Ltd., c1930, 3in (7.5cm) wide.
£7–8  *ATF*

*r.* A Speedway trophy,
1957, 6in (15cm) high.
£25–30  *COB*

No Time To Lose, by Rod Organ,
oil on canvas,
20 x 30in (50.5 x 76cm).
£900–1,000  *Rod*

A c1920 Agency for Triumph Motor
Cycles Enamel Sign.
£200–250  *BCA*

A Resin Figure riding on a Harley-
Davidson motorcycle, painted by Roy
Barrett, 19in (48cm) wide.
£75–80  *Bar*

A 1966 BSA Motorcycles Sign,
14in (35.5cm) wide.
£250–300  *DM*

A 1961 *Norton* Brochure,
7 x 9in (17.5 x 22.5cm).
£20–25  *DM*

An R.F. Motorcycle Helmet,
10in (25cm) wide.
£25–35  *COB*

# SUN *(British)*

**1956 Sun 224cc,** IH Cyclone Villiers 2-stroke engine.
**£600–800** *BLM*

**c1920 Sun** Villiers single cylinder 2-stroke.
**£1,500–2,000** *BLM*

# SUZUKI *(Japanese)*

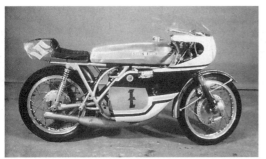

**1968 Suzuki TR250 Racer 247cc.**
**£4,000–5,000** *S*

**1968 Suzuki Super 6 246cc,** 2-stroke twin
cylinder, upswept exhaust, finished in red with
chrome and polished alloy, Swansea V5.
**Est. £2,000–2,700** *S*

**1968 Suzuki T20 Super 246cc,** in good original
condition, imported from USA.
**Est. £900–£1,000** *S*

**1971 Suzuki T125 Stinger 118cc,** twin cylinder
2-stroke engine with twin carburettors, good
condition and running order.
**Est. £300–500** *S*

*Some 4,000 examples were imported in 1971 and
the model is now rare.*

**1976 Suzuki TR750 Ex-Works 750cc,**
3-cylinder engine, close ratio gearbox, magnesium
carburettors and clutch cover, fully restored to
original specification, excellent condition.
**Est. £9,500–10,500** *BKS*

**1974 Suzuki TR500 Racing 492cc,** to factory
specification, in good running order.
**Est. £9,000–12,000** *BKS*

**c1976 Suzuki GT 250 247cc,** twin cylinder engine, 6-speed gearbox, complete and original.
**£300–400**  *S*

**1979 Suzuki RG500 Mk IV Racing,** good running order and condition throughout.
**£4,500–5,500**  *S*

# TRITON *(British)*

**1959 Triton 649cc.**
**£3,000–3,500**  *BLM*

*This red look-a-like Dominator had a T110 pre-unit engine with Norton gearbox.*

**1959 Triton 649cc,** with Norton wide-line frame, T110 engine restoration requiring completion.
**£1,600–1,800**  *MR*

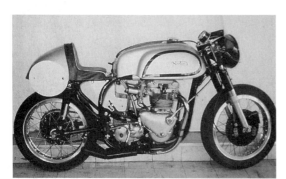

**1960 Triton 649cc,** alloy tank and rims, re-built to excellent condition.
**£2,200–2,500**  *MR*

**1960 Triton Special 500 Café Racer 499cc,** T100 engine, twin carburettors, twin leading shoe front brake, alloy rims, finished in black.
**£1,800–2,200**  *S*

> **Miller's is a price GUIDE not a price LIST**

**1960 Triton 649cc,** OHV vertical Triumph twin engine, hi-level exhaust, Norton featherbed frame, restored to excellent condition.
**£1,200–2,200**  *PS*

**1965 Triton 6T 649cc,** pre-unit Triumph engine, Norton Dominator frame.
**£3,000–3,400**  *BMM*

# TRIUMPH *(British 1902–)*

The famous Triumph marque was founded by two German immigrants, Siegried Bettmann and Maurice Schulte. They started manufacturing pedal cycles in Coventry in 1887.

It was not until 1905 when Triumph engaged Charles Hathaway to design a 300cc side valve single that they began to use their own engines. Hathaway followed up with 450 and 475cc versions. Sales rose rapidly from 500 in 1906 to over 3,000 in 1909 and a host of new models appeared from 1910–1920. The most successful of these was the 550cc side valve Model H, of which no less than 30,000 were supplied to the British Army. During the 1920s Triumph built the cheapest ever 500cc, the Model P, which sold for merely £43. So successful was this particular bike that production levels at one stage topped 1,000 a week.

Early in the 1920s, Triumph entered into car manufacturing, but this did not prove the wisest of moves. Not only was the car programme an expensive exercise, but the Wall Street crash of 1929 and its aftermath caused a massive slump in both two and four-wheel sales for the company. This so weakened Triumph that in 1936 it was sold to Jack Sangster who, through the design talents of Edward Turner, immediately set about rebuilding the motorcycle marque. Besides new singles of 250, 350 and 500cc, all of which featured OHV and chrome-plated tanks, Turner also created one of the most significant machines in motorcycle history, the 1937 Speed Twin. The new 499cc twin engine was in fact lighter than the same company's single cylinder Tiger 90 unit of the same period. It was a trend which was often copied, but rarely bettered by Triumph's rivals in succeeding years.

**1907 Triumph 3½hp,** good condition.
**Est. £4,000–5,000** *S*

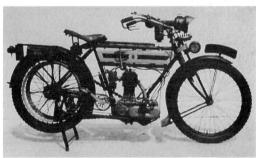

**1913 Triumph,** Sturmey-Archer 3-speed gearbox.
**£3,850–4,850** *VER*

**1915 Triumph Model H 550cc.**
**£4,000–4,500** *BLM*

**1918 Triumph Baby,** 2-stroke single cylinder engine, original unrestored condition.
**£2,500–3,000** *BCB*

*l.* **1920 Triumph H 550cc,** side valve single cylinder engine, 3-speed Sturmey-Archer gearbox with hand control, belt final drive, in Triumph livery, good condition, Swansea V5, VMCC Dating Certificate.
**£3,000–3,500** *S*

*One of the most famous British motorcycles of all time, the Model H was used extensively during WWI by the armed forces, and under the most arduous conditions proved itself to be both robust and reliable. It also saw extensive civilian service in the 1920s.*

**1921 Triumph Junior 225cc,** 2-stroke.
**£2,000–2,200** *BMM*

**1924 Triumph Ricardo 500cc,**
4-valve cylinder head.
**£5,000–6,000** *PVE*

**1925 Triumph SD,** 3-speed gearbox, chain drive,
finished in black with traditional silver, green and
red pinstriped petrol tank, good condition
throughout.
**£2,000–2,500** *S*

**1926 Triumph 2¾hp 350cc.**
**£3,000–3,500** *BLM*

**c1926 Triumph W 2¾hp 350cc.**
**£2,900–3,300** *BLM*

**1928 Triumph TT,** ex-works machine.
**£8,200–8,800** *VER*

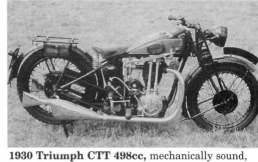

**1930 Triumph CTT 498cc,** mechanically sound,
in good condition.
**Est. £4,200–4,400** *BKS*

**1928 Triumph NSD 549cc,** side valve.
**£2,500–3,500** *BLM*
*Saddle tank model.*

> **Miller's is a price GUIDE
> not a price LIST**

*r.* **1934 Triumph XO5 148cc.**
**£1,500–2,000** *BLM*

**1935 Triumph L2/1 249cc,** twin port, OHV engine.
**£1,000–1,300** *BLM*

**1939 Triumph 3S 350cc,** side valve.
**£1,400–1,600** *BMM*

**1939 Triumph 3HW Grass Track Special 350cc,** OHV.
**£1,200–1,500** *BLM*

**1939 Triumph Tiger 80 343cc,** silver tank panels and high level competition exhaust, paintwork is in good condition.
**£3,000–4,000** *S*

**1940 Triumph Tiger 70 249cc,** single cylinder OHV 4-stroke, bore and stroke 63 x 80mm.
**£2,900–3,300** *C(A)*

**c1936 Triumph Tiger 70 249cc,** finished in silver and blue with chromium trim, in good condition.
**Est. £2,900–3,300** *S*

**1938 Triumph Tiger 80 343cc,** single cylinder OHV engine, girder front forks and rigid frame, in good condition.
**£1,500–2,000** *S*

Post-war Triumph became a major dollar earner with not only the latest version of the Speed Twin, but other models including the 649cc Thunderbird, 499cc Tiger 100 and its singles.

**1939 Triumph Tiger 100 499cc,** vertical twin, OHV, 4-stroke, 34bhp.
**Est. £3,500–5,000** *C(A)*

*This model is the sports version of the remarkable 5T Speed Twin and, as the name implies, it was capable of 100 mph.*

**1945 Triumph 3HW 350cc.**
**£1,800–2,000** *BLM*

**1947 Triumph Tiger 100 499cc,** OHV twin-cylinder, rigid frame, restored, in concours condition.
**£4,500–5,500** *RCB*

**1946 Triumph 3T 350cc,** panel tank, separate headlamp, rigid frame, superb example,.
**£3,800–4,000** *BLM*

**c1948 Triumph TR5 Trophy 499cc,** totally re-built in 1981 to original specification.
**£2,500–3,500** *MR*

**1948 Triumph 3T 350cc,** in good original condition.
**£1,600–2,000** *MR*

**1949 Triumph 5T Speed Twin 499cc,** sprung hub model.
**£3,200–3,600** *BLM*

**c1948 Triumph TR5 499cc,** totally re-built to original specifications.
**£2,500–3,500** *MR*

**1951 Triumph 6T Thunderbird 649cc,** sprung hub model.
**£3,000–3,500** *BLM*

**1950 Triumph TR5 499cc,** square barrelled with sprung hub, detachable lights for trials, 6:1 ratio.
**£2,000–3,000** *ST*

*r.* **1952 Triumph TR5 Trophy 500cc.**
**£2,500–3,500** *BMM*

**1953 Triumph Speed Twin 499cc,** OHV,
sprung hub.
**£3,500–4,250** *VER*

**1954 Triumph Tiger 100 499cc.**
**£4,500–5,000** *BLM*

*An early swinging arm Tiger.*

**1954 Triumph Tiger 100 499cc,** fully restored.
**£4,000–4,500** *TMSC*

*In 1954 Triumph introduced a new T100, the main
difference being a swinging arm frame and 8in
front brake.*

**1955 Triumph Thunderbird 6T 649cc,** swinging
arm rear suspension, coil ignition, alternator
instead of magneto and dynamo, restored to exact
original specification.
**Est. £2,800–3,200** *BKS*

*The Thunderbird was first introduced in 1949, with
a 649cc iron engine, and rigid or sprung hub rear.
It was fitted with a nacelle headlamp and
instrument fairing which was to set Triumph apart
from other manufacturers with its elegant styling,
complemented by striking polychromatic blue
paintwork and distinguished badging.*

**1955 Triumph Speed Twin 499cc,**
OHV twin-cylinder, restored to
concours condition.
**£2,900–3,500** *BCB*

**1956 Triumph TR5 499cc.**
**£3,750–4,750** *PCC*

**1956 Triumph Tiger 100 499cc,**
twin all-alloy engine, restored to
original condition.
**£3,500–4,000** *BCB*

**1956 Triumph TR5 Trophy 499cc.**
**£4,000–5,000** *BLM*

**c1958 Triumph 6T/TR6 649cc,** hybrid, mechanically sound.
**£2,000–2,200** *CStC*

**1958 Triumph Tiger 100 499cc.**
**£3,500–4,000** *BLM*

**1958 Triumph Tiger 100 499cc,** twin cylinder overhead valve, 4-speed gearbox, finished in black and cream livery, excellent condition, to original specification.
**Est. £3,000–3,500** *BKS*

**1959 Triumph 3TA Model Twenty One 349cc,** in T100A colours.
**£1,200–1,500** *BLM*

**1959 Triumph Model Twenty One 3TA 349cc.**
**£1,000–3,000** *BLM*

**1959 Triumph T110 Café Racer 649cc.**
**£1,500–2,000** *BLM*

**1959 Triumph T110 649cc.**
**£1,000–1,600** *PS*

**1959 Triumph TR6 Trophy 649cc,** finished in white and red, good mechanical condition.
**£5,000–6,000** *S*

*r.* **1959 Triumph T20 Tiger Cub T20 199cc.**
**£450–550** *AT*

**1960 Triumph TR20 Trials Tiger Cub 199cc.**
**£700–1,000** *BLM*

**1960s Triumph TRW 500cc,** side valve, twin,
ex-military model.
**£2,000–3,000** *BLM*

**1960 Triumph T120 Bonneville 649cc,**
with duplex frame.
**£4,800–5,000** *BLM*

Despite Triumph's success in the 1930s
and 1940s, the Company was sold to BSA
in 1951. As part of the BSA Group many
new models came on stream. The pick
being the Twenty-One, Bonneville and
Tiger Cub, at least from a marketing
point of view. Although Triumph
continued to meet sales success, other
parts of the BSA empire suffered badly
during the 1960s, resulting in the whole
group going bankrupt in 1973.

**1960 Triumph T120 Bonneville 649cc,** pre-unit,
completely re-built, Amal monobloc carburettors.
**£5,000–5,500** *TMSC*

**1961 Triumph T20 Tiger Cub 199cc,** dual seat,
sprung frame with telescopic forks, finished in
black with brown and gold tank.
**£450–550** *S*

**1961 Triumph T120 Bonneville 649cc,** imported
from USA, largely original, in good running order.
**Est. £2,500–2,800** *S*

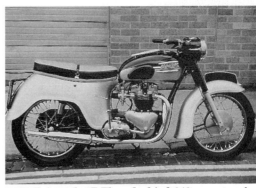

**1961 Triumph 6T Thunderbird 649cc,** restored,
in concours condition.
**£3,500–4,000** *BCB*

**c1962 Triumph TR20 Trial Cub.**
**£750–850** *BLM*

**1962 Triumph Twenty One 3TA 349cc,** twin
engine, 3-speed gearbox, in excellent condition.
**Est. £1,400–1,650** *S*

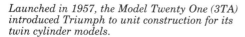

**1962 Triumph 3TA 349cc,** OHV twin
cylinder, restored.
**£1,500–1,750** *BCB*

*Launched in 1957, the Model Twenty One (3TA)*
*introduced Triumph to unit construction for its*
*twin cylinder models.*

**1963 Triumph Tiger 100SS 490cc.**
**£2,600–2,800** *BLM*

**1963 Triumph T120 Bonneville 649cc,** restored
to original specification, in concours condition,
finished in orange and white livery.
**Est. £4,200–4,600** *BKS*

**1963 Triumph 6T Thunderbird 649cc.**
**£2,000–3,000** *BLM*

**c1963 Triumph Trials 490cc.**
**£1,600–2,000** *SW*

**1963 Triumph T20 199cc.**
**£750–850** *PM*

*r.* **1963 Triumph Tiger 90,**
completely restored, all chromium
parts re-plated, finished in white.
**Est. £1,500–1,800** *S*

**1963 Triumph 5TA 490cc,** restored to original specification, finished in silver and black, good condition throughout.
**£3,500–4,000** *S*

**1964 Triumph 6T 649cc,** half 'bathtub' rear enclosure.
**£2,000–2,500** *PM*

**1964 Triumph Tiger 100SS 490cc,** OHV twin cylinder, re-built.
**£750–2,000** *BCB*

**1965 Triumph T20 Trials Cub 199cc,** not original.
**£1,000–1,200** *BMM*

**1965 Triumph Tiger 90 349cc.**
**£1,000–2,000** *PM*

**1965 Triumph 6T 649cc,** twin cylinder engine, 4-speed gearbox, complete mechanical re-build, good condition throughout, paintwork in metallic silver and black.
**£2,000–3,000** *S*

*l.* **1966 Triumph T120 Bonneville 649cc,** TLS front brake, export model with smaller fuel tank and no tank rack.
**£3,000–3,500** *BLM*

**1967 Triumph Tiger 90 349cc,** OHV unit construction, V-twin.
**£1,300–1,500** *PS*

*l.* **1967 Triumph T120 Bonneville 649cc,** restored to original condition, finished in burgundy and cream livery.
**£3,000–4,000** *BKS*

**1967 Triumph TR25 249cc,** OHV single, in good original condition.
**£700–950** *BCB*

**1967 Triumph Tiger 100C 490cc,** OHV twin cylinder, USA export specifications, in good original condition.
**£1,600–1,850** *BCB*

**1968 Triumph T100T.**
**£3,000–3,250** *VER*

**1968 Triumph T120 Bonneville 650cc,** V-twin, 4-speed gearbox, in good condition, paintwork in red and gold.
**Est. £3,000–3,500** *BKS*

**1968 Triumph Tiger 90 350cc,** in excellent restored condition.
**£1,000–1,500** *MR*

**1968 Triumph T120R,** restored to excellent condition, with Triumph certificate of authenticity.
**£3,800–4,200** *S*

*l.* **1968 Triumph Super Cub T20 199cc,** single cylinder.
**£700–800** *PM*

*This machine was commonly known as the Bantam Cub and was the last of the T20 line.*

**1969 Triumph T120 Bonneville 649cc,** OHV twin cylinder, restored to original condition.
**£3,250–3,750** *BCB*

**1969 Triumph T120 Bonneville 649cc,** OHV twin cylinder, re-built, in good condition.
**£3,250–3,750** *BCB*

**1970 Triumph TR6 Trophy 650cc.**
**£2,400–2,600** *BLM*

**1970 Triumph TR6 Trophy 650cc.**
**£3,000–4,000** *BLM*

**1970 Triumph T150 Trident**
**740cc,** USA model with small tank.
**£2,000–2,500** *PM*

**1971 Triumph T150 Trident**
**740cc,** OHV, 3-cylinder, re-built.
**£2,500–2,750** *BCB*

**1972 Triumph T25SS Blazer 247cc.**
**£1,000–1,200** *BLM*

**c1972 Triumph Trident Slippery Sam Replica,**
engine rebuilt.
**£2,500–2,700** *CStC*

**1973 Triumph T150 Trident 740cc.**
**£2,600–2,800** *BLM*

**1974 Triumph TR5 MX 499cc,** a rare
machine in original unrestored condition.
**Est. £1,700–2,000** *S*

*l.* **1974 Triumph T120 Bonneville 649cc,**
V-twin, 4-speed gearbox, in good condition,
finished in black and gold livery.
**£1,000–1,500** *BKS*

**1974 Triumph T140V Bonneville 744cc.**
OHV, twin cylinder, re-built.
**£2,000–2,500** *BCB*

**1975 Triumph T160 Slippery Sam
Replica 740cc.**
**£3,000–3,500** *BLM*

**1976 Triumph T160 Trident 740cc,**
3-cylinder engine, 5-speed gearbox, in excellent
condition to maker's specification, red paintwork.
**£5,200–5,800** *BKS*

**c1976 Triumph Bonneville T140 744cc,**
good condition.
**£1,500–1,700** *CStC*

In the 1980s and 1990s it was John Bloor
who kept the Triumph name alive. The
Bloor Triumphs with their modern 3 and
4-cylinders and water-cooling are
currently enjoying a sales boom, both in
the UK and overseas.

**1977 Triumph T140 Bonneville 744cc.**
**£2,700–3,000** *BLM*

**1977 Triumph T140 Bonneville 744cc.**
**£2,600–2,800** *BLM*

# VELOCE *(British)*

**1911 Veloce 499cc,** 3-cylinder, OHV
4-stroke, bore and stroke 67 x 70mm, 64bhp.
**£9,000–10,000** *C(A)*

*The forerunner to the legendary Velocette, this
model Veloce, was first introduced in 1910 and
is the only known surviving example in the
world. It originally sold for 40 guineas and was
exported new to Australia.*

**1977 Triumph Silver Jubilee Bonneville 744cc.**
**£2,000–3,000** *PM*

*This is number 130 of the special Silver
Jubilee model.*

# VELOCETTE *(British 1904–1968)*

Percy and Eugene Goodman were the men behind the creation of the Velocette marque. It was Percy Goodman's 1913 206cc 2-stroke single which firmly established Velocette on the motorcycling map. Velocette also went racing, first with tuned versions of its 2-stroke before becoming converts to the 4-stroke.

Sporting successes throughout the 1920s and 1930s did much to publicize Velocette motorcycles and sales of the production roadster flourished. Models such as the KTS and KSS overhead camshaft models were highly respected. Whilst the famous KTT was the customer version of the factory's race winning 350. In fact a Velocette was the first machine to win the 350cc World Championship, when Freddie Frith took the title in 1949 (Bob Foster continued the winning sequence the following year). During the 1950s the Hall Green, Birmingham

factory developed the 349cc Viper and 499cc Venom. In 1961 one of the latter machines, a Venom Clubman, with an Avon fairing and megaphone exhaust but otherwise standard, averaged over 100 mph at the Montlhéry circuit near Paris, for 24 hours, to set a new endurance record.

Viper and Venom machines did well in Sports Machine racing and then, in 1964 Velocette, introduced its ultimate sporting roadster single, the Thruxton.

Although its sports singles were revered by enthusiasts all around the world the company's management had made several bad decisions, notably with small capacity machines such as the Viceroy scooter and Vogue enclosed motorcycle. This, combined with relatively small sales of the expensive Thruxton, so weakened Velocette that it could not recover and they collapsed into liquidation in 1968.

**1923 Velocette G 249cc,** single cylinder 2-stroke engine, 3-speed gearbox, girder forks, rigid frame at the rear, excellent original condition, Swansea V5.
**£3,000–3,500** *S*

**1925 Velocette 250,** 249cc lightweight, 2-stroke, first with chain drive, unrestored, in original condition.
**£2,500–3,000** *Vel*

**1930s Velocette 250cc,** twin port.
**£2,000–3,000** *BLM*

**1934 Velocette KSS 348cc,** in good original condition.
**£2,500–3,500** *BKS*

**1935 Velocette Mk V KTT,** in good running order.
**Est. £10,000–12,000** *S*

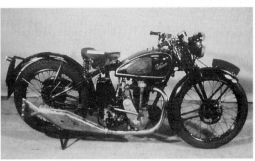

**1935 Velocette KSS Mk I 348cc.**
**£4,000–5,000** *S*

**1938 Velocette MAC 349cc,** single cylinder, OHV 4-stroke, bore and stroke 68 x 96mm, 15bhp, 4-speed foot change gearbox.
£4,500–5,500   *C(A)*

**1938 Velocette KSS MkII,** finished in traditional Velocette black and gold livery.
£4,000–4,500   *S*

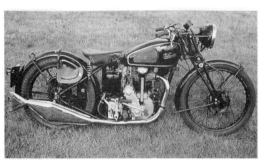

**1939 Velocette KSS 350cc.**
£3,800–4,200   *BKS*

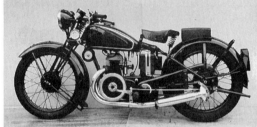

**1939 Velocette GTP 250cc,** 2-stroke single, in good overall condition.
£2,000–2,500   *VER*

**1947 Velocette MAC 349cc.**
£2,000–2,500   *PM*

**1939 Velocette Roarer 497cc.**
£250,000+   *VMCC*

*Built as a Grand Prix racer the famous Velocette Roarer was never used in earnest due to the advent of war and the subsequent ban on superchargers by the FIM.*

**c1947 Velocette KSS MkII 347cc,** OHC.
£4,500–5,000   *AtMC*

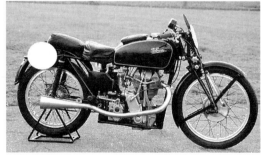

**1948 Velocette MkVIII KTT 348cc,** OHC, in excellent original condition throughout.
£10,500–11,000   *S*

*l.* **1949 Velocette MAC 349cc,** LHV, extensively restored, in good condition.
£2,200–2,500   *BKS*

**c1954 Velocette MAC 349cc,** pushrod OHV, mechanically re-built, Swansea V5 document.
**£1,700–2,500** *BKS*

**1954 Velocette MAC 349cc,** OHV, forerunner of Viper, restored.
**£2,500–3,000** *Vel*

**1955 Velocette MSS 499cc,** engine converted to Venom specification, sweptback exhaust pipe.
**Est. £2,300–2,800** *BKS*

**1955 Velocette MAC 350cc,** fully restored.
**£2,000–2,500** *MR*

**1955 Velocette LE200 192cc,** side valve horizontally opposed twin, in good original condition.
**£260–300** *PS*

**1956 Velocette MAC 349cc,** in good original condition, with maker's handbook and original buff log book.
**Est. £3,750–4,250** *S*

**1957 Velocette Venom 499cc,** to original specification, engine in good condition.
**£2,500–3,000** *S*

**1956 Velocette Viper Scrambler 349cc,** re-built, in excellent restored order, paintwork in black and gold.
**Est. £3,700–4,000** *S*

*This machine is believed to be one of only 25 built and was re-built by R. F. Seymour of Thame, Oxon, the Velocette specialists, in 1981.*

**1959 Velocette Venom 499cc.**
**£2,500–3,000** *PM*

**1959 Velocette MAC,** in good
original condition throughout.
**Est. £1,800–2,200** *S*

**1959 Velocette Viper 349cc,** single cylinder OHV
4-stroke, bore and stroke 72 x 86mm, 26bhp.
**£9,000–10,000** *C(A)*

*The Viper was Velocette's answer to the popular 350*
*BSA Goldstars.*

**1960 Velocette MAC 349cc.**
**£2,000–3,000** *BLM*

**1960 Velocette Venom 499cc.**
**£3,000–4,000** *BLM*

**c1960 Velocette Vogue 250cc.**
**£800–1,200** *BLM*

**c1960 Velocette Venom Special 499cc.**
**£2,500–3,500** *BLM*

**1960s Velocette Venom 499cc.**
**£3,000–3,500** *BLM*

*This is believed to be an ex-Earl's Court*
*Show model finished in powder blue.*

**Miller's is a price GUIDE**
**not a price LIST**

*l.* **1960 Velocette Valiant 192cc.**
**£1,000–1,500** *BLM*

**1960 Velocette Viper 349cc,** bore and stroke 72 x 86mm, producing 349cc, traditional black and chrome Velocette livery.
**£2,000–2,500** *BKS*

**1960 Velocette Viper 349cc,** swept back exhaust pipe.
**£1,500–2,100** *AT*

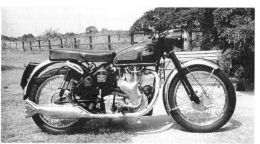

**1962 Velocette Venom 499cc.**
**£2,800–3,000** *BLM*

**1962 Velocette Venom Clubman 499cc.**
**£3,000–3,500** *BLM*

**c1964 Velocette Viper 349cc,** good original condition.
**£2,000–3,000** *SW*

**1963 Velocette MSS 499cc,** OHV, big single, painted black.
**£1,200–1,500** *S*

**1965 Velocette LE 192cc,** water-cooled.
**£300–350** *BMM*

**1965 Velocette Thruxton 499cc,** an excellent superbike classic, with original log book.
**Est. £6,900–7,200** *BKS*

# VERUS *(British)*

**1921 Verus 2¾hp 350cc,** Blackburne 4-stroke side valve single, Burman gearbox, chain-cum-belt drive, finished in black and blue livery.
**£2,500–3,000** *S*

*The Verus was built by Alfred Wiseman Ltd., of Glover Street, Birmingham, from 1919 until 1925. The Company also built the Sirrah and Weaver motorcycles at the cheaper end of the market.*

# VINCENT-HRD *(British 1924–1955)*

Founded in 1924 in Wolverhampton by TT winner Howard Raymond Davies, the original HRD Motors manufactured high quality sporting motorcycles using frames produced in-house and specially built JAP engines. The model 90 featured a single port 499cc JAP motor and the Super 90, a twin port racing JAP power plant with a maximum speed approaching the magic 100mph. All models featured hand change 3-speed Burman gearboxes.

HRD Motors went into receivership in 1927, but the machines continued to be manufactured until early 1928. The name and assets were purchased from the receiver by Ernie Humphries who then sold them on to Cambridge undergraduate, Philip C. Vincent. He had long been an enthusiast of HRD bikes and needed an established name for the motorcycles he was about to produce. These new bikes were Vincent-HRDs featuring Vincent's own patented rear suspension and using JAP and Rudge engines.

Vincent-HRD continued to use proprietary engines until a disastrous showing at the 1934 TT, when specially prepared JAP engines proved unreliable. This prompted Vincent to seek the services of the Australian engineer, Phil Irving, the pair then designed the 499cc engine for the Series A Vincent.

From this relationship was to stem such famous post-war models as the Comet and Grey Flash singles and, most of all, the now legendary Rapide, Black Shadow, Black Lightning and Black Prince 998cc OHV V-twins. Production ceased in 1955, but not the legend.

**1938 Vincent-HRD Series A Meteor 499cc,** re-built to specification and in concours condition. **£9,000–10,000** *BKS*

**1949 Vincent-HRD Rapide Series B 998cc,** extensively overhauled, in good condition. **Est. £11,500–12,000** *BKS*

**1950 Vincent Comet 499cc,** re-built and original log book. **Est. £5,000–5,500** *BKS*

**1949 Vincent-HRD Rapide 998cc,** V-twin OHV high cam 4-stroke, bore and stroke 84 x 90mm, 45bhp. **£10,000–12,000** *C(A)*

**1950 Vincent Series C Comet 499cc.** **£4,000–5,000** *BLM*

**1950 Vincent-HRD Series C Comet 499cc,** in original condition. **£3,500–3,700** *MR*

**1951 Vincent Series C Comet 499cc,** single cylinder OHV high cam 4-stroke, bore and stroke 84 x 90mm, 28.4bhp.
**£6,000–8,000** *C(A)*

**1951 Vincent Series C Grey Flash 499cc,** single cylinder OHV high cam 4-stroke, bore and stroke 84 x 90mm, 35bhp.
**£1,300– 2,000** *C(A)*

**1952 Vincent Rapide 998cc,**
in concours condition.
**£10,000–11,000** *BKS*

**1954 Vincent Series C Black Shadow 998cc,** V-twin OHV high cam 4-stroke, bore and stroke 84 x 90mm, 55bhp.
**£10,000–12,000** *C(A)*

*For many this is the classic Vincent. Sold as 'the world's fastest production motorcycle'. This sports version of the Rapide was capable of 125mph.*

# WANDERER *(German)*

**c1920 Wanderer 600cc,** V-twin engine with countershaft gearbox and Sturmey-Archer gear change lever, chain primary and final drive, fully restored, good condition.
**Est. £5,000–6,000** *BKS*

*This motorcycle is believed to be the only one of its type known in the UK.*

# WERNER *(French)*

**1904 Werner LD 249cc,** inlet over exhaust.
**£6,000–7,000** *AtMC*

# YAMAHA *(Japanese)*

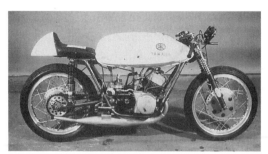

**1966 Yamaha TD1C 246cc.**
£8,000–10,000   *S*

**1967 Yamaha TD1C Racer 246cc,** air-cooled twin engine, 6-speed gearing, in full race trim, restored.
Est. £3,750–4,000   *S*

**1968 Yamaha Solo Racer 50cc.**
£3,000–4,000   *S*

*This 50cc Yamaha engined racer was built by Brian Woolley and raced by Trevor Burgess.*

**1969 Yamaha/Bultaco TD2 Racer 125cc,** combination of Bultaco frame with Yamaha water-cooled R5 race engine.
Est. £2,800–3,200   *BKS*

**1971 Yamaha TR2B Racing 348cc,** air-cooled engine, full race trim, in good original unrestored condition.
£2,800–3,200   *BKS*

**1971 Yamaha TR2B Racing 348cc,** ready to race, with documented modifications, finished in red and white
Est. £3,500–4,000   *S*

**1976 Yamaha TZ750 Racing 750cc,** totally stripped and re-built to works racing specification.
Est. £8,000–9,000   *BKS*

**1973 Yamaha TA125 Racing 124cc,** water-cooled, full race fairing, restored.
Est. £2,000–2,250   *S*

**1974 Yamaha TZ Racing 750cc,** water-cooled 4-cylinder engine, in full race trim, excellently restored.
**£7,000–8,000** *S*

**1978 Yamaha/Maxton TZ750 Racing,** water-cooled, 4-cylinder race prepared engine, in excellent condition.
**Est. £5,800–6,200** *BKS*

*l.* **1978 Yamaha XS750,** DOHC.
**£850–1,000** *S*

**1979 Yamaha/Spondon TZ750,** water-cooled 4-cylinder race engine.
**£3,500–4,500** *BKS*

*Spondon Engineering have been the makers of quality and innovative conversions for race machines for many years with great success and are indeed still in demand today.*

## ZENITH *(British)*

**1920s Zenith 800cc,** V-twin.
**£6,000–7,000** *BLM*

## ZUNDAPP *(German)*

**1958 Zündapp 201S 198cc,** complete and original.
**£50–60** *S*

*The German Zündapp factory, at Nurenberg, and later Munich, from 1921 produced 2-stroke machines exclusively for the first 10 years. Before, during and after WWII they made large transverse flat twins in various sizes, but towards the end of the 1950s they reverted to more modest 2-stroke engined motorcycles and scooters.*

# SPECIALS

**1970 Aermacchi/Hartman Special, Switzerland
246cc,** Ala Verde engine, special bodywork and
square section swing arm.
**£2,000–3,000** *PC*

**1967 AJS Special 500cc.
£3,000–5,000** *BLM*

*This unique street scrambler was commission-built
and has a rather special frame.*

**AJS Special Racing,** rolling chassis, AJS duplex
frame, forks fitted with Royal Enfield twin-sided
braked front hub, AMC rear hub and racing
gearbox, no engine available.
**Est. £1,000–1,200** *BKS*

*The chassis was commenced many years ago but is
still an unfinished project.*

**1972 Mick Walker Ducati 340cc.
£3,500–4,500** *PC*

*One of 25 special racing machines constructed by
Mick Walker, using either Saxon or Spondon
frames, from 1972–74.*

**1960 Matchless G80 Special 497cc,** includes
Norton forks, front wheel, alloy rims, 7R seat, clip-
on handlebars, rear-sets and Gold Star silencer.
**Est. £2,500–3,000** *S*

*This is a Clubman's special machine.*

**c1961 Matchless/Norton Special 350cc,**
slim-line frame, G3L engine and AMC gearbox.
**£1,700–2,200** *BLM*

**1971 Simpson/Norton F750 Road Racer 750cc,**
commando engine, Quaif 5-speed gearbox, Norvil
front disc brake, Seeley type frame.
**£5,500–6,000** *HOC*

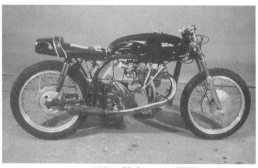

**1950s Velocette KTT 250 248cc,**
Norton forks and front wheel.
**£6,000–7,000** *S*

# SIDECARS

**1920 AJS Model D 770cc Combination,**
V-twin, A. J. Stevens 3-speed countershaft
gearbox, engine rebuilt.
**£6,500–7,500** *BKS*

*This machine has regularly rallied in the UK,
continental events and the Dutch Veteran Rally
at Hengelo.*

**c1970 AJS/Otter 360cc.**
**£700–900** *BLM*

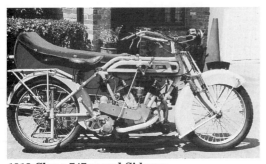

**1913 Clyno 747cc and Sidecar.**
**£8,500–9,000** *VER*

**c1932 Harley-Davidson Model U
1300cc Combination,** side valve
engine, 3-speed gearbox.
**Est. £6,000–7,000** *BKS*

**1926 AJS Model G 7.99hp Combination,** restored
to include a new Alpha big end, new valves, in
concours condition.
**£6,000–7,000** *BKS*

*Best machine overall at the VMCC International West
Kent run 1986 and 1992.*

**c1960 BMW R50/Steib sidecar 494cc.**
**£5,500–6,500** *AtMC*

**1961 BSA A10 Gold Flash Solo Combination.**
**Est. £2,700–3,000** *CStC*

*This A10 Gold Flash has been fitted with a 1980s
Squire single seat sidecar.*

**c1966 Honda 49cc Monkey Bike with
Sidecar,** good condition.
**£1,500–2,000** *S*

*This bike was originally supplied by Honda UK
to Arthur Boddey for his son in recognition of
the fact that he was one of the first Honda 'five
star' dealers.*

**1915 Indian 990cc Combination,** overhead inlet valves, leaf spring suspension front and rear.
**£9,500–10,500** *S*

**1921 Matchless S8 998cc Combination.**
**£6,000–7,000** *S*

**1956 Norton 19S/Steib 597cc Combination,** restored to an excellent standard.
**£3,500–4,500** *S*

**1949 Moto Guzzi Falcone 500cc Combination,** in good running order.
**Est. £6,500–7,500** *S*

*This combination features a 2-seater sidecar by Longhi, including a baby seat.*

**1933 OEC 1150cc and Torpedo Sidecar by Watsonian,** JAP LTZ side valve V-twin engine, Burman gearbox, duplex steering, in black and silvery liver, in concours condition.
**Est. £10,000–10,500** *BKS*

**c1918 Panther 3½hp Combination,** in good running order.
**£4,500–5,000** *S*

*Jonah Phelon and Harry Rayner built the first P & M motorcycle in Cleckheaton in 1900 and were joined by Richard Moore, the bikes taking their name from Phelon and Moore and later adopting the name Panther.*

*l.* **1912 Rover 3½hp Combination,** wicker Canoelet sidecar by Mead & Deakin.
**£7,000–7,850** *VER*

**1920s Sunbeam 550cc Combination,** fully restored and in concourse condition.
**£4,000–5,000** *BCB*

**1923 Sunbeam 3½hp 500cc Combination,** side valve, John Marston Ltd. (Wolverhampton) Sunbeam chair.
**Est. £4,800–5,200** *BKS*

# SCOOTERS

**1946 Corgi 98cc,** Excelsior Spryt
engine, 2-speeds and chain drive.
**Est. £1,200–1,500**  *S*

**1949 Corgi 98cc,** Excelsior Spryt horizontal
2-stroke engine, folding handlebars.
**£350–400**  *S*

**1960 Garelli Capri 79cc,**
finished in 2-tone grey and white.
**£35–45**  *S*

**1959 Dolphin 98cc,** Villiers
2-stroke engine, unrestored.
**£60–80**  *S*

**1961 James 150 Scooter 149cc,**
AMC 15H 2-stroke.
**£500–600**  *PM*

**1921 Kingsbury 125cc.**
**£2,200–2,800**  *S*

*After WWII Kingsbury Aviation Company turned to
making the Kingsbury Junior light car, the
Kingsbury motorcycle and a motor scooter, and they
flourished briefly in this period.*

**1959 Lambretta Li Series 2, 150cc,** twin
individual seats, finished in white and black.
**£60–100**  *S*

**1954 Lambretta D150 148cc.**
**£800–1,200**  *EP*

**1960 Lambretta Li 150 Series 2 150cc,**
unrestored.
£60–120  *S*

**1960 Lambretta Li 150 Series 2,** twin
seats, finished in 2-tone red and white
and in good condition.
£600–700  *S*

**1969 Lambretta SX200 198cc,** restored.
Est. £1,100–1,200  *PS*

**1953 MV Agusta Ovunque 149cc,**
good condition.
£350–400  *BKS*

**1960 Trobike Solo,** Clinton engine.
£200–250  *ALC*

*The Trobike was produced in the 1960s by
Lambretta Concessionaires and could have been
said to bridge the gap between the wartime
Corgi and 1970s Honda 'monkey' bike. Perhaps
the answer to today's traffic problems!*

**1965 Triumph T10.**
£600–800  *BLM*

**1957 Zündapp Bella R154 198cc,**
dual pillion, old style log book.
£600–500  *BKS*

*This machine has been laid up in dry
storage since 1977.*

**1960 Zündapp R204 Bella 198cc.**
£600–700  *BMM*

# MOPEDS

**1955 Heinkel Perle 50cc.**
**£400–500**  *BMM*

**1981 Honda Caren 49cc,** believed to be in good
condition, with a Swansea V5.
**Est. £350–400**  *S*

*This machine was produced for use with the
female shopper in mind.*

**1970 Mobylette Minor 49cc,** this French-built
moped has automatic transmission and centrifugal
clutch, front wheel needs replacing, tyres poor
otherwise in good condition.
**Est. £80–100**  *S*

*This moped has had one owner and covered
only 640 miles, the engine is free and turning.
A straightforward restoration project.*

**1972 Mobylette 50cc.**
**£100–150**  *S*

**1958 NSU Quickly,** leading link front end, rigid
rear, dual seat, finished in red and blue, Swansea V5.
**£20–100**  *S*

**1961 NSU Quickly Special,** good order
throughout, 4,320 miles believed to be genuine,
2 owners only since 1983.
**£180–300**  *BKS*

**1958 Raleigh,** in need of restoration,
no registration documents.
**£12–20**  *BKS*

**1965 NSU Quickly 49cc,** 2-stroke,
3-speed, original condition.
**Est. £150–175**  *PS*

# MINI MOTORCYCLES

**1974 Fantic Collapsible Mini Motorcycle 50cc,** chopper handlebars, gold and chrome trim, good condition.
**£350–450** *S*

*An Italian collapsible version of the Japanese Monkey Bike.*

**1969 Honda Monkey Bike 50cc,** genuine 600 miles recorded, good condition, finished in red.
**£1,000–1,200** *S*

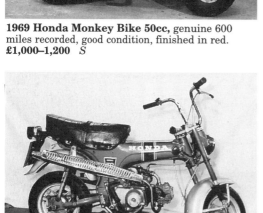

**1976 Honda ST70,** 72cc engine.
**£250–350** *PS*

**1978 Honda ST70 Dax 72cc,** air-cooled engine, semi-automatic clutch, folding handlebars.
**£350–450** *CARS*

**1965 Honda Monkey Bike 49cc,** good condition throughout.
**£2,000–2,500** *S*

*With one owner since 1975 this bike has been dry-stored in a garage ever since.*

**1971 Honda Mini Trail Motorcycle 49cc,** 176 miles recorded, very good condition, finished in yellow.
**£570–700** *BKS*

**1977 Honda ST70,** 72cc engine.
**£500–600** *S*

**1980 Suzuki RV125,** a detuned version of the TS trail bike engine, 5-speed gearbox, complete, with Swansea V5.
**£800–1,000** *S*

# RESTORATION PROJECTS

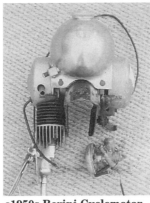

**Vincent Firefly Under-The-Frame Mounting Cycle Motor 48cc,** complete with petrol tank, ancillaries, mounting brackets, engagement mechanism and controls.
**£270–350** *S*

**A British Salmson Aero Engines Cyclaid Cyclemotor,** over rear wheel mounted and belt driven, complete with fuel tank, ancillaries, mounting brackets, rear wheel drive hoop and controls, engine no. 1947.
**£350–400** *BKS*

**c1950s Berini Cyclemotor,** engine suitable for front mounting, complete with tank, exhaust and ancillaries, engine no. 72180.
**£110–150** *BKS*

**c1950s Vincent Firefly 48cc Cyclemotor,** suitable for under frame mounting, with fuel tank, ancillaries, mounting brackets, engagement mechanism and controls, engine no. TO5AB1-51021.
**£470–550** *BKS*

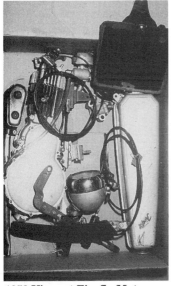

**JAP Side Valve Static 499cc Engine,** complete with Amal carburettor and starting handle.
**Est. £100–150** *S*

> **Miller's is a price GUIDE not a price LIST**

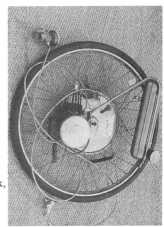

**1953 Vincent Firefly Motor,** and ancillaries.
**£300–350** *S*

*r.* **BSA Winged Wheel 32cc Engine,** suitable for rear mounting, complete with tank, carburettor and exhaust, engine no. MBW.12011.
**£300–400** *BKS*

**1960 James Captain 200,** 197cc 2-stroke AMC engine, complete.
**Est. £400–600** *S*

**c1966 BSA A65 Hornet 654cc,** American import, good project.
**£850–950** *CStC*

**Cymotor Front Mounting Engine,** complete with exhaust, carburettor and cowl, with integral headlamp and grille.
**£500–600** *BKS*

**Trojan Mini Motor 49cc Engine,** in unused condition, complete, in original box with instruction manual, engine no. B33198.
**£350–450** *BKS*

**1953 Vincent Black Knight Rolling Chassis,** a nearly complete Series D rolling chassis.
**Est. £3,000–4,000** *S*

*This machine would be ideal for an enthusiast with the time and patience to search out the missing components.*

**1924 Norton 633cc,** original.
**£1,500–2,000** *VMCC*

**1948 NSU 250cc,** pre-WWII engine in post-war frame, not original.
**£250–300** *BCB*

**1935 Norton Model 16H 490cc,** carburettor missing, with old style buff log book and Swansea V5.
**£870–1,000** *BKS*

**c1925 Triumph Model Q 500cc,** single cylinder side valve engine, 3-speed countershaft gearbox, original and unrestored.
**£950–1,250** *BKS*

**Triton 499cc,** Triumph T100 pre-unit engine, gearbox and dismantled clutch, Norton wideline frame and hubs with alloy rims, fibreglass fuel tank and seat, clip-on handlebars, unfinished project.
**£600–1,000** *BKS*

**c1924 Ravat Type ER 98cc,** air-cooled, 2-stroke, Longuemare carburettor and magneto ignition, pedal assistance.
**£500–700** *BKS*

# MOTORCYCLE MEMORABILIA

## Signs

A 1930s Castrol Motor Oil Enamel Sign.
£100–140  *BCA*

A 1920s Royal Enfield Bicycles Enamel Sign.
£80–100  *BCA*
A 1940s Bougies Eyquem Thermometer Sign.
£70–90  *BCA*

A 1930s 'BP' Sold Here Double-sided Enamel Sign, poor condition.
£100–150  *MSMP*

A Shell Economy Petrol Globe, white opaque glass.
£150–170  *MR*

A 1950s Shell Aviation Double-sided Enamel Sign, with flange mounting, very good condition.
£300+  *MSMP*

A 1966 Triumph Motorcycles Sign, 14 x 6 x 8in (35.5 x 15 x 20cm).
£150–175  *DM*

A 1950s Norton, light box, in good condition.
£170–200  *MSMP*

## Programmes & Books

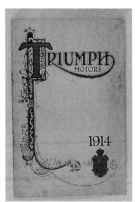

**Triumph Motors 1914,**
catalogue,
9 x 16in (22.5 x 40.5cm).
**£80–100**  *DM*

**James Motorcycles**
**1922,** 9½ x 6in
(24 x 15cm).
**£40–45**  *DM*

**The Ariel Modern**
**Cycle,** 1931 brochure,
6 x 9½in (15 x 24cm).
**£30–35**  *DM*

**The MotorCycle,** 1929
magazine,
11 x 8in (28 x 20cm).
**£10–15**  *DM*

*l.* **BSA The**
**Most Popular**
**Motor Cycle in**
**the World,** 1955
brochure,
10 x 7in
(25 x 17.5cm)
**£15–20**  *DM*

**A Terrot Colour**
**Lithograph Poster,**
on linen, 28½ x 30½in
(72.5 x 77cm).
**£250–280**  *ONS*

**Matchless Motorcycles**
**1935,** brochure,
10 x 7in (25 x 17.5cm).
**£25–30**  *DM*

# Motorcycle Art

**The Foxy Lady, by Roy Barrett**
Signed, watercolour, 16½ x 12½in (41.5 x 32cm).
**£250–300** *Bar*

**The Firebird by Roy Barrett**
Signed, watercolour, 16½ x 12½in (41.5 x 32cm).
**£250–300** *Bar*

**Champions, by Rod Organ**
Signed, oil on canvas,
24 x 36in (61 x 91.5cm).
**£1,000–1,500** *Rod*

**Great Scot, by Rod Organ**
Signed, oil on canvas,
24 x 33in (61 x 83.5cm).
**£800–900** *Rod*

## Miscellaneous

**Two Rubber Magnets.**
£1–5 each  *COB*

**1909 Douglas Motors Gold Medal,** 1¼in (3.5cm) diam.
**£800–1,000+**  *LDM*

**1914 Motor Cycling Club Silver Medal,** 1¼in diam.
**£20–30**  *LDM*

*Given to P. Phillips for riding in London / Lands End, 11-13th April 1914.*

**1955 Sunbeam MCC Moonbeam Run Trophy Ashtray,** 4¾in (12cm) diam.
**£10–12**  *COB*

*l.* **An early Motorcycle Helmet,** 9½in (24cm) long.
**£25–28**  *COB*

**Motorcycle Helmet,** tweed peak, 11in (28cm) long.
**£30–35**  *COB*

*r.* **A Lucas Oil Side Light,** brass, 8¾in (22cm) high.
**£40–50**  *ONS*

## Petrol Pumps

**c1920 Bonniksen Motorcycle Speedometer,** by Rotherhams of Coventry.
**£450–500**  *HOLL*

**1960s Shell Petrol Pump.**
**£150–200**  *BMM*

**A Shell Motor Spirit Pump,** hand petrol pump, with Shell globe, restored.
**£550–600**  *ALC*

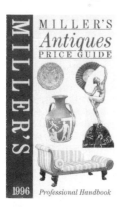  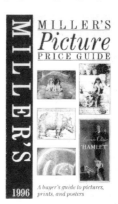
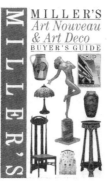 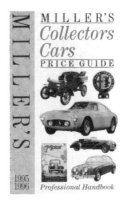 

166

# GLOSSARY

*We have attempted to define some of the terms that you will come across in this book. If there are any other terms or technicalities you would like explained or you feel should be included in future editions, please let us know.*

**ACU** - Auto Cycle Union, who control a large part of British motorcycle sport.

**Advanced ignition** - Ignition timing set causing firing before the piston reaches centre top, variation is now automatic.

**Air-cooling** - Most motorcycles rely on air-cooling to the atmosphere.

**Air intake** - The carburettor port admitting air to mix with fuel from the float chamber.

**AMCA** - Amateur Motor Cycle Association, promoters of English off-road events.

**APMC** - The Association of Pioneer Motor Cyclists.

**Auto Cycle Club** - Formed in 1903 it was the original governing body of motorcycle sport, in 1907 became the Auto Cycle Union.

**Automatic inlet valve** - Activated by the engine suction. Forerunner of the mechanically operated valve.

**Balloon tyres** - Wide section, low pressure, soft running tyres, used on tourers for comfort.

**Beaded-edge tyres** - Encased rubber beads in channel on wheel rim.

**Belt drive** - A leather or fabric belt from engine or gearbox to rear wheel.

**BHP** - A measure of engine output, e.g. to lift 33,000lb one foot in a minute requires one horsepower.

**BMCRC** - British Motor Cycle Racing Club, formed in 1909.

**BMF** - British Motorcycle Federation.

**Bore/stroke ratio** - Cylinder diameter ratio to stroke.

**Cam** - Device for opening and closing a valve.

**Camshaft** - The mounting shaft for the cam, can be in low, high or overhead position.

**Carburettor** - Used to produce the air/fuel mixture.

**Chain drive** - Primary form of drive from engine to gearbox and secondary gearbox to rear wheel.

**Combustion chamber** - Area where the fuel/air mixture is compressed and fired, between piston and cylinder head.

**Compression ratio** - The fuel/air mixture compression degree.

**Crankcase** - The casing enclosing the crankshaft and its attachments.

**Crankshaft** - The shaft for converting the up-and-down piston motion into rotary.

**Cylinder** - Containing the piston and capped by the cylinder head, is the site of the explosion which provides power.

**Cylinder head** - In a vertical engine caps off the top end of the cylinder. In a 4 stroke engine carries the valves.

**Damper** - Used for slowing down movement in suspension system or as crankshaft balance.

**Displacement** - The engine capacity or amount of volume displaced by the movement of the piston from bottom dead centre to top dead centre.

**Distributor** - A gear driven contact sending high tension current to spark plugs.

**DOHC** - Double overhead camshaft.

**Dry sump** - Two oil pumps, one supplying oil to the bearings from a tank, the other to return it to the tank.

**Earles forks** - An unusual front fork design. A long leading link and rigid pivot through both links behind the wheel.

**Featherbed** - A Norton frame, designed by Rex and Crommie McCandless, Belfast, used for racing machines from 1950, road machines from 1953.

**FIM** - Federation Internationale Motorcycliste, controls motorcycle sport worldwide.

**Flat head** - A flat surfaced cylinder head.

**Flat twin** - An engine with 2 horizontally opposed cylinders, or 4 to make a Flat Four.

**Float** - A plastic or brass box which floats upon the fuel in a float chamber and operates the needle valve controlling the fuel.

**Flywheel** - Attached to the crankshaft this heavy wheel smooths intermittent firing impulses and helps slow running.

**Friction drive** - An early form of drive using discs in contact instead of chains and gears.

**Gearbox** - Cased trains of pinion wheels which can be moved to provide alternative ratios.

**Gear ratios** - Differential rates of speed between sets of pinions to provide higher or lower rotation of the rear wheel in relation to the engine.

**GP** - Grand Prix, an international race to a fixed formula.

**High camshaft** - Mounted high up on the engine to shorten the pushrods in an ohv formation.

**IOE** - Inlet over exhaust, a common arrangement with an overhead inlet and side exhaust.

**Leaf spring** - Metal blades clamped and bolted together, used in suspension many years ago.

**Magneto** - A high tension dynamo producing current for the ignition spark. Superseded by coil ignition.

**Main bearings** - Bearings in which the crankshaft runs.

**Manifold** - Collection of pipes supplying mixture or taking away fumes.

**MCC** - The Motor Cycling club which runs sporting events. Formed in 1902.

**Moped** - A light motorcycle of under 50cc with pedals attached.

**OHC** - Overhead camshaft, can be either single or double.

**OHV** - Overhead valve engine.

**Overhead cam** - An engine with overhead camshaft or camshafts operating its valves.

**Overhead valve** - A valve mounted in the cylinder head.

**Pinking** - A distinctive noise from an engine with over-advanced ignition or inferior fuel.

**Piston** - The component driven down the cylinder by expanding gases.

**Post-vintage** - A motorcycle made after December 31, 1930 and before January 1, 1945.

**Pressure plate** - The plate against which the clutch springs react to load the friction plates.

**Pushrods** - Operating rods for overhead valves, working from cams below the cylinder.

**Rotary valve** - A valve driven from the camshaft for inlet or exhaust and usually a disc or cylinder shape. For either 2 or 4 stroke engines.

**SACU** - Scottish Auto Cycle Union, which controls motorcyle sport in Scotland.

**SAE** - Society of Automotive Engineers. Used in a system of classifying engine oils such as SAE30, IOW/50 etc.

**Shock absorber** - A damper, used to control up-and-down movement of suspension or to cushion a drive train.

**Silencer** - Device fitted to the exhaust system of an engine whereby the pressure of the exhaust gases is considerably reduced before reaching the outer air.

**Swinging arm** - Rear suspension by radius arms carrying the wheel and attached to the frame at the other end.

**Torque** - Twisting rotational force in a shaft, can be measured to show at what point an engine develops most torque.

# INDEX TO ADVERTISERS

# BIBLIOGRAPHY

Bacon, Roy; British Motorcycles of the 1930s, Osprey, 1986.
Bacon, Roy; Matchless & AJS Restoration, Osprey, 1993.
Bacon, Roy; Norton Twin Restoration, Osprey, 1993.
Bacon, Roy; Triumph Twins & Triples, Osprey, 1990.
Birkitt, Malcolm; Harley-Davidson, Osprey, 1993.
Champ, Robert Cordon; Sunbeam S7/S8 Super Profile, Haynes, 1983.
Davis, Ivor; It's a Triumph, Haynes, 1980.
Morley, Don; and Woollett, Mick; Classic Motorcycles, BMW, Osprey, 1992.
Morley, Don; Classic Motorcycles, Triumph, Osprey, 1991.
Stuart, Garry; and Carroll, John; Classic Motorcycles, Indian, Osprey, 1994.
Tragatsch, Erwin, ed; The New Illustrated Encyclopedia of Motorcycles, Grange Books, 1993.

Walker, Mick; Classic Motorcycles, Honda, Osprey, 1993.
Walker, Mick; Classic European Racing Motorcycles, Osprey, 1992.
Walker, Mick; Classic Italian Racing Motorcycles, Osprey, 1991.
Walker, Mick; Classic Japanese Racing Motorcycles, Osprey, 1991.
Walker, Mick; Classic Motorcycles, Ducati, Osprey, 1993.
Walker, Mick; Classic Motorcycles, Kawasaki, Osprey, 1993.
Walker, Mick; Classic Motorcycles, Suzuki, Osprey, 1993.
Walker, Mick; Classic Motorcycles, Yamaha, Osprey, 1993.
Wherrett, Duncan; Classic Motorcycles, Vincent, Osprey, 1994.
Woollett, Mick; Norton, Osprey, 1992.

# DIRECTORY OF MOTORCYCLE CLUBS

*If you wish to be included in next year's directory or if you have a change of address or telephone number, please could you inform us by April 30th 1996. Entries will be repeated in subsequent editions unless we are requested otherwise.*

**AJS & Matchless Owners Club,**
25 Victoria Street, Irthlingborough, Northants NN9 5RG
Tel: 01933 652155

**AMC Owners Club,**
c/o Terry Corley, 12 Chilworth Gardens, Sutton,
Surrey SM1 3SP

**Androd Classics,**
70 Broadway, Frome, Somerset, Avon BA11 3HE
Tel: 01373 471087

**Ariel Owners Club,**
c/o Mike Taylor, Harrow House, Woolscott, Rugby,
Warwickshire CV23 8DB

**Ariel Owners Motor Cycle Club,**
Swindon Branch, 45 Wheeler Avenue, Swindon,
Wiltshire SN2 6HQ

**Ariel Owners Motor Cycle Club,**
c/o Lester Grant, Cedar Cottage, Dicklow Cob,
Lower Withington, Macclesfield, Cheshire SK11 9EA

**Bantam Enthusiasts Club,**
c/o Vic Salmon, 16 Oakhurst Close, Walderslade,
Chatham, Kent ME5 9AN

**Benelli Owners Club,**
c/o Vic Salmon, 16 Oakhurst Close, Walderslade,
Chatham, Kent ME5 9AN

**Benelli Owners Club,**
c/o Rosie Marston, 14 Rufford Close, Barton, Seagrave,
Kettering, Northamptonshire NN15 6RF

**BMW Club,**
c/o John Lawes (Vintage Secretary), Bowbury House, Kirk,
Langley, Ashbourne, Derbyshire DE6 4NJ
Tel: 01332 824334

**BMW Owners Club,**
c/o Mike Cox, 22 Combermere, Thornbury,
Bristol, Avon BS12 2ET
Tel/Fax: 01454 415358

**Bristol & Avon Roadrunners Motorcycle Club,**
177 Speedwell Road, Speedwell, Bristol, Avon BS5 7SP

**Bristol & District Sidecar Club,**
158 Fairlyn Drive, Kingswood, Bristol, Avon BS15 4PZ

**Bristol Genesis Motorcycle Club,**
Burrington, 1a Bampton Close, Headley Park, Bristol,
Avon BS13 7QZ
Tel: 0117 978 2584

**British Motor Bike Owners Club,**
c/o Ray Peacock, Crown Inn, Shelfanger, Diss,
Norfolk IP22 2DL

**British Motorcycle Owners,**
c/o Phil Coventry, 59 Mackenzie Street, Bolton,
Lancashire BL1 6QP

**British Motorcyclists Federation,**
129 Seaforth Avenue, Motspur Park, New Malden,
Surrey KT3 6JU

**British Two Stroke Owners Club,**
c/o Mark Hathaway, 45 Moores Hill, Olney,
Bucks MK46 5DY

**Brough Superior Club,**
c/o Piers Otley, 6 Canning Road, Felpham,
Sussex PO22 7AD

**BSA Owners Club,**
44 Froxfield Road, West Leigh, Havant, Hants PO9 5PW

**CBX Riders Club,**
c/o Peter Broad, 57 Osborne Close, Basingstoke,
Hampshire RG21 2TS

**Chiltern Vehicle Preservation Group,**
Chiltern House, Aylesbury, Buckinghamshire HP17 8BY.
Tel: 01296 651283

**Christian Motorcycle Association North,**
c/o Mr A. Sutton, 100 Low Bank Road, Ashton-in-
Makerfield, Wigan, Greater Manchester WN4 9RZ

**Classic Motorcycle Racing Club,**
c/o Simon Wilson, 6 Pendennis Road, Freshwook,
Swindon, Wiltshire SN5 8QD
Tel: 01793 610828

**Cossack Owners Club,** c/o Mr Charles Hancock, Lake
View, Carr Road, North Kelsey, Lincolnshire LN7 6LB

**Cotton Owners & Enthusiasts Club,**
c/o Peter Turner, Coombehayes, Sidmouth Road,
Lyme Regis, Dorset DT7 3EQ

**DKW Rotary Owners Club,**
c/o David Cameron, Dunbar, Ingatestone Road,
Highwood, Chelmsford, Essex CM1 3QU

**Dot Owners Club,**
c/o Chris Black, 115 Lincoln Avenue, Clayton,
Newcastle-upon-Tyne, Tyne & Wear ST5 3AR

**London Douglas Motorcycle Club,**
c/o Reg Holmes (Membership Secretary),
48 Standish Avenue, Stoke Lodge, Patchway, Bristol

**Ducati Owners Club,**
131 Desmond Drive, Old Catton, Norwich, NR6 7JR

**Featherbed Specials Owners Club,**
Maytham Farm, Maytham Road, Rolvenden, Cranbrook,
Kent TN17 4NP

**Francis Barnett Owners Club,**
58 Knowle Road, Totterdown, Bristol, Avon BS4 2ED

**Gold Star Owners Club,**
c/o George Chiswell, 43 Church Lane,
Kitts Green, Birmingham, West Midlands B33 9EG

**Goldwing Owners Club,**
82 Farley Close, Little Stoke, Bristol, Avon BS12 6HG

**Greeves Owners Club,**
c/o Dave McGregor, 4 Longshaw Close, North Wingfield,
Chesterfield, Derbyshire S42 5QR

**Greeves Riders Association,**
40 Swallow Park, Thornbury, Bristol, Avon BS12 1LS
Tel: 01454 418037

**Harley Davidson Owners Club,**
1 St Johns Road, Clifton, Bristol, Avon BS8 2ET

**Harley Davidson Riders Club of Great Britain,**
SAE to Membership Secretary, PO Box 62, Newton Abbott,
Devon TQ12 2QE

**Hesketh Owners Club,**
c/o Tom Wilson, 19 Stonnall Road, Aldridge, Walsall,
West Midlands W59 8JX

**Historic Raleigh Motorcycle Club,**
c/o R. Thomas, 22 Valley Road, Solihull,
West Midlands B92 9AD

**Honda Owners Club (GB),**
c/o Dave Barton, 18a Embley Close, Calmore,
Southampton, Hampshire SO40 2QX

**Indian Motorcycle Club,**
c/o John Chatterton (Membership Secretary),
183 Buxton Road, Newtown, New Mills, Stockport
Tel: 01663 747106

**International CBX Owners Association,**
24 Pevensey Way, Paddock Hill, Frimley, Camberley,
Surrey GU16 5YJ
Tel: 01252 836698

**International Laverda Owners Club,**
c/o Alan Cudipp, Orchard Cottage, Orchard Terrace,
Acomb, Hexham, Northumberland NE46 4QB

**Italian Motorcycle Owners Club,**
c/o Rosie Marston (Membership Secretary),
14 Rufford Close, Barton Seagrave, Kettering

**Jawa-CZ Owners MCC,**
c/o Peter Edwards, 2 Churchill Close, Breaston,
Derbyshire DE72 3UD

**Kawasaki Owners Club,** c/o John Dalton,
37 Hinton Road, Runcorn, Cheshire WA7 5PZ

**L E Velo Club,**
c/o Peter Greaves, 8 Heath Close, Walsall,
West Midlands WS9 9HU

**Laverda Owners Club,** c/o Ray Sheepwash,
8 Maple Close, Swanley, Kent BR8 7YN

**Le Velocette Club,**
32 Mackie Avenue, Filton, Bristol, Avon BS12 7ND

**Maico Owners Club,**
c/o Phil Hingston, No Elms, Goosey, Nr Faringdon,
Oxfordshire FN7 8PA

**Military Vehicle Trust,**
PO Box 6, Fleet, Hampshire GU13 9PE

**Morini Owners Club,**
c/o Richard Laughton, 20 Fairford Close, Church Hill,
Redditch, Hereford & Worcester B98 9LU

**Morini Riders Club,**
c/o Kevin Bennett, 1 Glebe Farm Cottages, Sutton Veney,
Warminster, Wiltshire BA12 7AS

**Moto Guzzi Club GB,**
c/o Jenny Trengove, 53 Torbay Road, Harrow,
Middlesex HA2 9QQ

**MV Agusta Club GB,**
c/o Martyn Simpkins, 31 Baker Street, Stapenhill,
Burton-on-Trent, Staffordshire DE15 9AF

**MV Agusta Owners Club,**
c/o Ray Gascoine, 7 Lowes Lane, Wellisbourne,
Nr Warwick, Staffordshire CV35 9RB
**MZ Riders Club (South West),**
c/o Alex Pearce, 80 Kingskirswell Road, Newton Abbott,
Devon TQ12 1DG
Tel: 01626 331584
**National Autocycle & Cyclemotor Club,**
c/o Rob Harknett, 1 Parkfields, Roydon, Harlow,
Essex CM19 5JA
**National Hill Climb Association,**
43 Tyler Close, Hanham, Bristol, Avon BS15 3RG
Tel: 0117 944 3569
**New Imperial Owners Association,**
c/o Mike Slater, 3 Fairview Drive, Higham, Kent ME3 7BG
**North Devon British Motorcycle Owners Club,**
c/o Mrs Y Coleman, Bassett Lodge, Pollards Hill,
Little Torrington, Devon EX00 0JA
**Norton Owners Club,**
c/o Dave Fenner, Beeches, Durley Brook Road, Durley,
Southampton, Hants SO32 2AR
**Panther Owners Club,**
c/o A & J Jones, Coopers Cottage, Park Lane, Castle
Camps, Cambridge,Cambridgeshire CB1 6SR
**Rickman Owners Club,**
c/o Michael Foulds, 35 Otterbourne Road, Chingford,
London E4 6LL
**Riders for Health,** The Old Vicarage, Norton,
Nr Daventry, Northamptonshire NN11 5ND
**Royal Enfield Owners Club,**
c/o John Cherry, Meadow Lodge Farm, The Hollows,
Coalpit Heath, Bristol, Avon BS17 2UX
**Rudge Enthusiasts Club,**
c/o Colin Kirkwood, 41 Rectory Green, Beckenham,
Kent BR3 4HX
Tel: 0181 658 0494
**Scott Owners Club,**
c/o H Beal, 2 Whiteshott, Basildon, Essex SS16 5HF
**Shrivenham Motorcycle Club,**
12-14 Townsend Road, Shrivenham, Swindon, Wilts SN6 8AS
**Street Specials Motorcycle Club,**
55 Haldon Close, Bedminster, Bristol, Avon BS3 5LR
**Street Specials Motorcycle Club,**
c/o E Warrington, 8 The Gallops, Norton, Malton,
Yorkshire YO17 9JU

**Sunbeam Owners Club,** c/o Stewart Engineering,
Church Terrace, Harbury, Leamington Spa, Warwickshire
CV33 9HL
**Sunbeam Owners Fellowship,**
PO Box 7, Market Harborough, Leicestershire LE16
**Suzuki Owners Club,** Mark Fitz-Gibbon, 3 Rossetti
Lodge, Burns Road, Royston, Hertfordshire SG8 5SF
**Sidecar Register,** c/o John Proctor, 112 Briarlyn Road,
Birchencliffe, Huddersfield, Yorkshire HD3 3NW
**Trident and Rocket Three Owners Club,**
63 Dunbar Road, Southport, Merseyside PR8 4RJ
**Triumph Motorcycle Club,** 6 Hortham Lane,
Almondsbury, Bristol, Avon BS12 4JH
**Triumph Owners Club,** c/o Mrs M Mellish, 4 Douglas
Avenue, Harold Wood, Romford, Essex RM3 0UT
**Velocette Owners Club,**
c/o David Allcock, 3 Beverley Drive, Trinity Fields,
Stafford, Staffordshire SR16 1RR
**Velocette Owners Club,**
c/o Vic Blackman, 1 Mayfair, Tilehurst, Reading,
Berkshire RG30 4RA
**Vespa Club of Great Britain,**
c/o Mr S Barbour (Membership Secretary), 254 Braehead,
Bonhill, Alexandria, Dunbartonshire, Scotland G83 9NE
**Vincent Owners Club,**
c/o Andy Davenport, Ashley Cottage, 133 Bath Road,
Atworth, Wiltshire SN12 8LA
**Vintage Japanese MCC,**
c/o John Dalton, 1 Maple Avenue, Burchill,
Onchan Douglas, Isle of Man IM3 3HG
**Vintage Japanese Motorcycle Club,** 9 Somerset
Crescent, Melksham, Wiltshire SN12 7LX
Tel: 01225 702816
**Vintage Motor Scooter Club,**
c/o Ian Harrop, 11 Ivanhoe Avenue, Lowton St Lukes,
Nr Warrington, Cheshire WA3 2HX
**Vintage Motor Cycle Club,**
Allen House, Wetmore Road, Burton-on-Trent,
Staffordshire DE14 1TR
Tel: 01283 540557 Fax: 01283 510547
**Vintage Motorcycle Club of Ulster,** c/o Mrs M Burns,
20 Coach Road, Comber, Newtownards, Co Down,
Ireland BT23 5QX
**ZI Owners Club,** c/o Sam Holt, 54 Hawhome Close,
Congleton, Cheshire CW12 4UF

# DIRECTORY OF MUSEUMS

**Battlesbridge Motorcycle Museum**
Muggeridge Farm, Maltings Road, Battlesbridge,
SS11 7RF.
Tel: 01268 769392/560866
An interesting collection of classic motorcycles &
scooters in a small informal museum. Open Suns
10.30am-4pm. Adults £1, children free.

**Birmingham Museum of Science & Industry**
136 Newhall Street, Birmingham, B3 1RZ.
Tel: 0121 235 1651
A small collection of motorcycles. Open Mon to Sat
9.30am-5pm. Sun 2pm-5pm. Closed December 25-26,
and January 1. Admission free.

**Bristol Industrial Museum**
Princes Wharf, City Docks, Bristol, BS1 4RN.
Tel: 0117 925 1470
A small collection of Bristol-made Douglas machines,
including the only surviving V4 of 1908, and a 1972
Quasar. Open Saturday to Wednesday 10am-1pm
and 2pm-5pm. Closed Thurs and Fri, also Good
Friday, December 25-27 and January 1. Adults £2,
under 16s free.

**Brooklands Museum**
The Clubhouse, Brooklands Road, Weybridge, KT13
0QN. Tel: 01932 857381
The birthplace of British motorsport and aviation,
Brooklands has several museums on display. Open
Saturday and Sunday 10am-4pm. Guided tours at
10.30am and 2pm on Tues, Weds and Thurs. Adults
£4, OAPs & students £3, children £2.

**Cotton's Classic Bikes, Phil**
Victoria Road Museum, Ulverston, LA12 0BY.
Tel: 01229 586099
Working museum, most exhibits are available to buy.
Open 10am-4.30pm Tues-Sat, closed Sun & Mon.

**Craven Collection of Classic Motorcycles**
Brockfield Villa, Stockton-on-the-Forest, YO3 9UE.
Tel: 01904 488461/400493
Private collection of over 180 Vintage & Post-War
Classic Motorcycles. Open to the public on first
Sunday of every month and Bank Holiday Mondays,
10am-4pm. Club visits & private parties arranged.
Adults £2.00, Children under 10 Free.

**Foulkes-Halbard of Filching**
Filching Manor, Jevington Road, Wannock, Polegate,
BN26 5QA. Tel: 01323 487838
30 motorcycles, including pre-'40s American bikes
ex-Steve McQueen, 100 cars, 1893-1993. Open 7 days a
week in summer 10.30-4.30pm. Thurs-Sun in winter,
or by appt. Adults £3, OAPs and children £2.

**Grampian Transport Museum**
Main Street, Alford, Aberdeenshire, AB33 8AD.
Tel: 019755 62292
30-40 machines ranging from a 1902 Beeston
Humber to a Norton F1. Mods and Rockers caff
display with Triton and Triumph Tina scooter.
Competition section with 1913 Indian twin and 1976
Rob North replica Trident racer. Open March 28-
October 31, 10am-5pm. Adults £2.30, children 80p,
OAPs £1.50, family ticket £5.

**Haynes Sparkford Motor Museum**
Sparkford, Yeovil, BA22 7LH. Tel: 01963 440804
30 plus machines from 1914 BSA onwards. Video
theatre. Bookshop. Open Mon-Sun 9.30am-5.30pm.
Closed December 25-26 and January 1. Adults £4.50,
OAPs £4, children £2.75.

**Murray's Motorcycle Museum**
Bungalow Corner, TT Course, Isle of Man.
Tel: 01624 861719
140 machines, with Hailwood's 250cc Mondial and
Honda 125cc and the amazing 500cc 4 cylinder
roadster designed by John Wooler. Open May-Sept
10am-5pm. Adults £2, OAPs & children £1.

**Museum of British Road Transport**
St. Agnes Lane, Hales Street, Coventry, CV1 1PN.
Tel: 01203 832425
65 motorcycles, from local firms such as Coventry
Eagle, Coventry Victor, Francis-Barnett, Triumph
and Rudge. Close to city centre. Open every day
except December 24-26, 10am-5pm. Adults £2.50,
children, OAPs and unemployed £1.50.

**Museum of Transport**
Kelvin Hall, 1 Bunhouse Road, Glasgow, G3 8DP.
Tel: 0141 357 3929
A small collection of motorcycles including
Automobile Association BSA combination. Open
Mon-Sat 10am-5pm. Sunday 11am-5pm. Closed
December 25 and January 1. Admission free.

**Myreton Motor Museum**
Aberlady, Longniddry, East Lothian, EH32 0PZ.
Tel: 018757 288
A small collection including 1926 350cc Chater-Lea
racer and Egli Vincent. Open Easter to Oct 10am-
5pm and Oct to Easter 10am-6pm. Closed Dec 25
and Jan 1. Adults £2, children 50p.

**National Motor Museum**
Brockenhurst, Beaulieu, SO42 7ZN.
Tel: 01590 612123/612345
Important motorcycle collection. Reference and
photographic libraries. Open Easter to Sept 10am-
6pm, Oct to Easter 10am-5pm. Closed Dec 25.
Adults £6.75, OAPs/students £5.25, children £4.75
(includes Museum, rides and drives, Monastic Life
Exhibition and entry to Palace House and grounds).

**National Motorcycle Museum**
Coventry Road, Bickenhill, Solihull, B92 0EJ.
Tel: 01675 53311

**Royal Museum of Scotland**
Chambers Street, Edinburgh, EH1 1JF.
Tel: 0131 225 7534
Small display of engines and complete machines
including the world's first 4 cylinder motorcycle, an
1895 Holden. Open Mon to Sat 10am-5pm. Sun 2pm-
5pm. Closed Dec 25, Jan 1. Admission free.

**Sammy Miller Museum,**
Gore Road, New Milton, BH25 6TF.
Tel: 01425 619696
Sammy Miller is a living legend in the world of
motorcycle racing, and the museum was opened in
1983 by John Surtees. All bikes are in working order
and wherever possible are run in classic bike events.
At present there are 200 bikes in the Museum, many
extremely rare. New exhibits are being sought all
the time, with much of the restoration work being
carried out on the premises by Sammy Miller
himself. There are interesting artefacts and items of
memorabilia, including many cups and trophies won
by Sammy over the years. A typical motorcycle
workshop of 1925 has been reconstructed. Open
10.30am-4.30pm every day, April-Oct 10.30am-
4.30pm, Sats and Suns Nov-March. 15 miles west of
Southampton and 10 miles east of Bournemouth at
New Milton, Hants.

**Science Museum**
Exhibition Road, South Kensington, SW7 2DD.
Tel: 0171 589 3456
Collection of engines and complete machines, with
cutaway BSA A10, Yamaha XS1100, 1940 500cc
BMW and 1969 Honda CB750. Mon-Sat 10am-6pm.
Sun 11am-6pm. Closed Dec 24-26. Adults £4, OAPs and
children £2.10, disabled free. Most Science Museum's
motorcycle collection is at Wroughton Airfield near
Swindon, Wilts. Tel: 0793 814466.

**Stanford Hall Motorcycle Museum**
Stanford Hall, Lutterworth, LE17 6DH.
Tel: 01788 860250
Older machines and racers. Open Sats, Suns,
Bank Holiday Mondays and following Tues, Easter
to Sept, 2.30pm-6pm. (12 noon-6pm when a special
event is taking place.). Admission to grounds:
Adults £1.60, children 70p. Museum: Adults 90p,
children 20p.

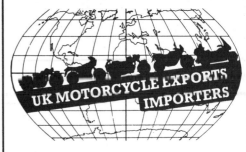

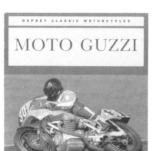
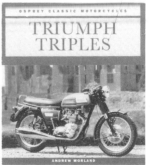

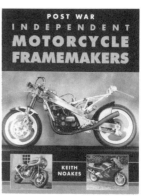